HOW TO POWER TUNE
MGB
4-CYLINDER ENGINES
FOR ROAD & TRACK

NEW UPDATED & EXPANDED EDITION

PETER BURGESS

VELOCE PUBLISHING
THE PUBLISHER OF FINE AUTOMOTIVE BOOKS

– In memory of Wilson McComb and Bill Nicholson –

SpeedPro Series

4-Cylinder Engine Short Block High-Performance Manual – New Updated & Revised Edition (Hammill)
Aerodynamics of Your Road Car, Modifying the (Edgar and Barnard)
Alfa Romeo DOHC High-performance Manual (Kartalamakis)
Alfa Romeo V6 Engine High-performance Manual (Kartalamakis)
BMC 998cc A-series Engine, How to Power Tune (Hammill)
1275cc A-series High-performance Manual (Hammill)
Camshafts – How to Choose & Time Them For Maximum Power (Hammill)
Competition Car Datalogging Manual, The (Templeman)
Custom Air Suspension – How to install air suspension in your road car – on a budget! (Edgar)
Cylinder Heads, How to Build, Modify & Power Tune – Updated & Revised Edition (Burgess & Gollan)
Distributor-type Ignition Systems, How to Build & Power Tune – New 3rd Edition (Hammill)
Fast Road Car, How to Plan and Build – Revised & Updated Colour New Edition (Stapleton)
Ford SOHC 'Pinto' & Sierra Cosworth DOHC Engines, How to Power Tune – Updated & Enlarged Edition (Hammill)
Ford V8, How to Power Tune Small Block Engines (Hammill)
Harley-Davidson Evolution Engines, How to Build & Power Tune (Hammill)
Holley Carburetors, How to Build & Power Tune – Revised & Updated Edition (Hammill)
Honda Civic Type R High-Performance Manual, The (Cowland & Clifford)
Jaguar XK Engines, How to Power Tune – Revised & Updated Colour Edition (Hammill)
Land Rover Discovery, Defender & Range Rover – How to Modify Coil Sprung Models for High Performance & Off-Road Action (Hosier)
MG Midget & Austin-Healey Sprite, How to Power Tune – Enlarged & updated 4th Edition (Stapleton)
MGB 4-cylinder Engine, How to Power Tune (Burgess)
MGB V8 Power, How to Give Your – Third Colour Edition (Williams)
MGB, MGC & MGB V8, How to Improve – New 2nd Edition (Williams)
Mini Engines, How to Power Tune On a Small Budget – Colour Edition (Hammill)
Motorcycle-engined Racing Cars, How to Build (Pashley)
Motorsport, Getting Started in (Collins)
Nissan GT-R High-performance Manual, The (Gorodji)
Nitrous Oxide High-performance Manual, The (Langfield)
Race & Trackday Driving Techniques (Hornsey)
Retro or classic car for high performance, How to modify your (Stapleton)
Rover V8 Engines, How to Power Tune (Hammill)
Secrets of Speed – Today's techniques for 4-stroke engine blueprinting & tuning (Swager)
Sportscar & Kitcar Suspension & Brakes, How to Build & Modify – Revised 3rd Edition (Hammill)
SU Carburettor High-performance Manual (Hammill)
Successful Low-Cost Rally Car, How to Build a (Young)
Suzuki 4x4, How to Modify For Serious Off-road Action (Richardson)
Tiger Avon Sportscar, How to Build Your Own – Updated & Revised 2nd Edition (Dudley)
Triumph TR2, 3 & TR4, How to Improve (Williams)
Triumph TR5, 250 & TR6, How to Improve (Williams)
Triumph TR7 & TR8, How to Improve (Williams)
V8 Engine, How to Build a Short Block For High Performance (Hammill)
Volkswagen Beetle Suspension, Brakes & Chassis, How to Modify For High Performance (Hale)
Volkswagen Bus Suspension, Brakes & Chassis for High Performance, How to Modify – Updated & Enlarged New Edition (Hale)
Weber DCOE, & Dellorto DHLA Carburetors, How to Build & Power Tune – 3rd Edition (Hammill)

www.veloce.co.uk

First edition published in 1996; reprinted 1997, 1999, 2001, 2002, 2003, and 2006. New Updated & Revised Edition August 2013, reprinted June 2014 and April 2018 by Veloce Publishing Limited, Veloce House, Parkway Farm Business Park, Middle Farm Way, Poundbury, Dorchester, Dorset, DT1 3AR, England.
Tel: 01305 260068/Fax 01305 250479/e-mail info@veloce.co.uk/web www.veloce.co.uk or www.velocebooks.com.

ISBN: 978-1-787113-41-1 UPC: 6-36847-01341-7

Contents

SPEEDPRO SERIES

Thanks & about the author

My thanks to David Gollan B ENG HONS for doing the lion's share of the work, including the photographs and line drawings; Rog Parker for his knowledge of suspension and brakes, and Robert Day for photo developing and printing and for teaching us to use a camera!

Also, Terry Hurrell at Moss London; Kash and the lads at Moss Oldbury; Gerry Brown; Alan Scott; Doug Smith; Shaun Powell; Martin Hardy; Max Smith; Alan Peacock; Michael Lee; Keith Peats, Nick Walker and Phil Gollan.

Keith Hick of the MG Owners' Club was kind enough to vet and approve the original manuscript.

For this updated edition, a big thank you to Keith Hippey and Jordan Dent, and to Lloyd Faust and Mike Ellsmore for letting us know of misprints and errors.

PETER BURGESS

Peter was born in Byfleet in Surrey about a mile from the Brooklands race circuit.

He has been modifying MGBs since 1978 and runs a small, friendly tuning business which enjoys worldwide distribution.

Peter's technical and theoretical knowledge of engines and horsepower production has resulted in many MGB race-wins and Championships in Circuit Racing, Sprints, Hill Climbs, Trialling and Rallying every year from 1991 to this completely revised edition in 2013.

Drivers using Peter's cylinder heads and engines range from absolute novices taking to the track for the first time, to legendary drivers such as Stirling Moss.

Essential information & using this book

ESSENTIAL INFORMATION

This book contains information on practical procedures; however, this information is intended only for those with the qualifications, experience, tools and facilities to carry out the work in safety and with appropriately high levels of skill. Remember that your personal safety must **ALWAYS** be your **FIRST** consideration.

The publisher, author, editors and retailer of this book cannot accept any responsibility for personal injury or mechanical damage which results from using this book, even if caused by errors or omissions in the information given. If this disclaimer is unacceptable to you, please return the pristine book to your retailer who will refund the purchase price.

This book applies specifically to MGB & MGB GT cars with four-cylinder (5-main bearing) B-series engines, though it will also prove useful in tuning any BMC/British Leyland B-series engine.

Please be aware that changing the engine's specification may mean that it no longer complies with exhaust emission control regulations in your state or country – check before you start work.

You need to know that varying production tolerances in cylinder head castings can mean that breakthrough to the waterways or other passages can occur if metal is removed by grinding. If such breakthrough does occur it's most likely that the head will have to be scrapped.

An increase in engine power and therefore performance will mean that your car's braking and suspension systems will need to be kept in perfect condition and uprated as appropriate.

USING THIS BOOK

Throughout this book the text assumes that you or your contractor will have an appropriate MGB workshop manual to follow for complete detail on dismantling, reassembly, adjustment procedure, clearances, torque figures, etc. This book's default is the standard manufacturer's specification, so if a procedure is not described, a measurement not given, a torque figure ignored, assume that the manufacturer's procedure or specification for your car should be used.

You'll find it helpful to read the whole book before you start work or give instructions to your contractor. This is because a modification or change in specification in one area can cause the need for changes in other areas. Get the whole picture so that you can finalize specification and component requirements as far as possible before any work begins.

You'll find the term "thou" turning up quite often in the text, it simply means one thousandth of an inch (0.001in/ 0.0254mm).

Introduction

The aim of this book is to convey information based on my experience of tuning MG four-cylinder engines, and to give an overview of the type and level of tuning options currently available. You should then be able to make a more informed choice in order to improve your car's performance.

Topics are covered in sufficient depth to interest readers who have already spent many years improving their MGBs, whilst endeavouring to introduce performance tuning to others who drive a standard MGB and are not familiar with, or have previously been put off by, the technical language of 'tuners.'

When performance tuning an engine, components are modified or changed in order to improve function and efficiency, in turn improving efficiency of various engine processes. With the modern computer-designed and controlled types of engine this is becoming more and more difficult to achieve. The optimisation process

has, to a large extent, already been completed by the manufacturer during initial design and development, but financial constraints, legislation and market forces still limit the extent of factory performance tuning. Older designs such as the B series, however, offer far more scope for tuning and modification. Years of revision and refinement by the aftermarket tuning industry have allowed considerable performance gains to be made; gains often far in excess of those achieved by the early factory tuning programmes.

Much the same can be said about performance tuning the car's suspension and braking systems. While, again, modern cars are becoming more and more sophisticated in terms of ride, handling and stopping power, any driving sensations or feedback have been lost in favour of comfort and refinement. Hopping in your favourite classic for a drive can sometimes come as a bit of a culture shock, albeit a very pleasant

one, after a period of driving a modern clone. Here again, upgrades to improve performance in this department are available.

Given the age of some MG four-cylinder cars, it's not surprising that some components have begun to reach the end of their useful life, so maybe now is the time to investigate the options for upgrading to parts that have been modified and improved, or which enjoy the benefits of more modern design and manufacturing methods.

It takes a long time to understand an engine fully and know how it responds to various component changes – this also applies to suspension and brakes. Due to the myriad of options and mass of often conflicting advice available nowadays, converting your car can be an extremely daunting, not to mention potentially expensive, experience. The problem is how to go about sorting out what you want from all that information ...?

First and most important is to decide exactly what you want from the car in terms of performance. Formula One cornering with acceleration to match is all very well but not much use if your car is mostly used in traffic. All you may really want is a little more sparkle in the engine performance department, or more mid-range power for laid-back effortless cruising: either way, be honest with yourself, and, once decided on a plan of campaign, stick to it.

It can be very beneficial to talk to other owners (don't forget racers, sprinters and hillclimbers) who have tried, or are still using, modifications or tuning firms that you may be thinking of trying yourself. They are usually more than happy to share experiences as to which of the many tuners can offer a good level of service as well as quality products. Couple that with the information given in this book and you can then go out and achieve the results you want.

We begin with a brief description of the standard 1800 B series engine, its history and the various production changes it underwent. It is an intentionally brief section as the history and development of the cars is already very well documented elsewhere. The second chapter explains the fundamentals of the four stroke cycle, as a lead into covering terms such as torque, horsepower, volumetric efficiency and various other technical parameters, cornerstone definitions for the testing and improvement of internal combustion engines.

Chapter three offers a guide to cost-effective engine tuning and which modifications represent the best value for your hard-earned money. These have been assessed from the viewpoint of which components offer the biggest overall power gain, down to those that, while they may not contribute a significant power increase, go a long way towards improving engine performance and reliability.

Chapter four contains both the original power recipes, which list a variety of different component combinations and the various power outputs achieved from them, and additional new recipes together with corresponding power and torque curves obtained from our new rolling road.

Cylinder heads are covered in great detail in chapter five, offering very comprehensive descriptions and explanations. Included are converting to run unleaded fuel; modifications ranging from mild to wild; the gains that can achieved by straightforward DIY work done at home, right up to the more serious changes that make up a professional full race conversion.

Chapter six covers engine building, large bore conversions and blueprinting techniques. It shows machining and assembly techniques for improving reliability, as well as wringing the most horsepower out of the engine – methods that can be applied to even the most basic freshen-up rebuild.

The ups and downs of camshaft selection occupy chapter seven, covering camshaft terminology, the effects of cam selection on torque and horsepower, how to correctly install and bed-in a cam, and any necessary valvetrain considerations.

Carburation – getting the air and fuel mixed in the correct proportions and then into the engine – makes up chapter eight, with advice on carburettor selection for different applications, together with the necessary inlet manifolds and essential air filters and other fuel system components.

Setting the mixture alight is a task performed by the ignition system, covered in chapter nine, including ignition requirements for standard and modified engines, and taking in such topics as distributors, advance curves, the pros and cons of vacuum advance, plug leads and sparkplugs. The benefits of electronic ignition systems are also mentioned.

Chapter ten on exhaust manifolds and systems concludes the mechanical side of the power plant, covering some of the various types of exhaust system available and their effects on performance.

Chapter eleven on lubrication and cooling offers a guide to ensuring the engine continues to live long and prosper in either standard or modified form.

Transmission and suspension chapters twelve and thirteen describe alternative gearboxes, axles and tyres and suspension and braking conversions to ensure that the extra engine power unleashed is matched by the car's ability to take bends and stop safely!

Chapter fourteen covers what has to be the most vital part of getting the best from any engine – rolling road tuning – and offers a step-by-step guide to setting up a standard car during a rolling road session to ensure it gives its best, including practical advice for setting up carburettors and ignition. The latter can be of benefit to anyone at home with a few basic tools and some spare time. We have included an example modern dyno tuning session from our new rolling road, as we are now able to show graphs of power and torque for each step of the process.

Chapter fifteen offers some suggestions toward achieving improved fuel economy. With fuel prices spiralling ever upwards we decided to add a few thoughts on getting more miles per gallon.

The Appendix contains the maths and formulae more commonly used in performance tuning, including some examples and a list of suppliers for the components covered in this book.

While methods of forced induction such as turbocharging or supercharging can contribute to significant performance gains, it was decided to concentrate on the subject of tuning normally-aspirated engines.

Chapter 1
Engine

The forerunner of the 1800 MGB engine started life as a 1200cc power unit originally fitted to the Austin A40 Cambridge.

The BMC B-series 4-cylinder engine was first fitted as a production unit to the MGA in the early 1950s. Whilst derived from the smaller engine, a redesign meant capacity had grown to 1489cc by increasing the bore and stroke, which changed again in 1959 to give 1588cc. The latter replaced the double overhead cam unit of the MGA twin cam with a more reliable pushrod unit, its capacity increased through a larger bore size. The final version of the engine used for the MGA was 1622cc; the capacity increase again by virtue of a larger cylinder bore size. This unit also differed in that the cylinder head benefited from larger valves. In fact, the cylinder head fitted to the last MGAs was virtually identical to that fitted to the first MGBs in 1962. For the MGB, the engine was again increased in capacity, to 1798cc.

About the only thing in the engine that didn't change during production for MGA and MGB sports cars was the camshaft specification, which produced a good broad spread of usable power in all the engine's various guises. Incidentally, with a change of part number (to 88G229), the same cam profile also ran very nicely in the 997cc Mini Cooper engine!

The MGA and early MGB engines (18G and 18GA) had three main bearing bottom ends and tended to suffer from crankshaft whip and vibration problems. These were addressed for the MGB around the time of the company's rationalisation of engine manufacture, with the next generation of engines (18GB) improved by the use of a redesigned block and crank that had five main bearings, originally developed to power the then new Austin 1800.

The next step forward (18GD to 18GG) came with the change to pistons that had three piston rings (top, second

and oil control) instead of the four (top, second and two oil control rings) used in earlier engines, which allowed power improvements through reduced friction. The design of the connecting rods was also changed, to one with a horizontal parting line for the bearing cap in place of the earlier angled type.

The final derivation of the MGB engine (18V) came into production in 1971. Whilst the block remained the same, the cylinder head was revised and used a kidney-shaped combustion chamber in place of the more compact, heart-shape of earlier engines. The inlet valve was enlarged in size from 1.567 inches (39.8mm) to 1.625 inches (41.3mm) diameter. In order to maintain a similar compression ratio, the chambers in the 18V heads were shallower (by some 40 thousandths of an inch), and the physical thickness of the head was reduced. This necessitated further changes to shorter cam followers and pushrods in order to re-establish the correct valvetrain geometry (using the earlier longer pushrods and followers made it impossible to set the tappet clearances, as the tappets ran out of adjustment).

There was also a second, and more serious, problem with the 18V. Due to the different head thickness, the exhaust valve almost hit the top of the block when at full lift (contact guaranteed during valve bounce!). To overcome this the 18V blocks had small cutouts machined into the top of the bores to allow the exhaust valves to safely enter the bore and eliminate the chances of disastrous contact.

The 18V engine used the same camshaft as its predecessors but its timing was 4 degrees more advanced, which didn't help the engine rev well or make good top end power, offering instead reasonable low rpm torque and cleaner idling emissions.

With bigger inlet valves it would be reasonable to expect more power but this turns out not to be the case. Both the smaller valved 18G engines and the big valve 18V engines produce the same horsepower and torque!

Unfortunately the rubber bumper Bs had cylinder heads fitted with smaller inlet valves – back to the 1.56 inch

(39.8mm) diameter of the 1622 MGA and 18G MGB engines. This markedly reduced the revability of the engine and it became very sluggish above 4000rpm. The smaller valve did, however, give an increase in mid-range torque, which was intended to help move the now much heavier car.

A table of typical power outputs for the different standard engine specifications, as measured on a rolling road, is given below.

	MGA 1500	MGA 1600	MGA 1600 MkII	MGB 18G	MGB 18V	Rubber bumper MGB 18V 74 on
Engine size	1498cc	1588cc	1622cc	1798cc	1798cc	1798cc
Bore and stroke	70.03/88.9	75.42/88.9	76.2/88.9	80.26/88.9	80.26/88.9	80.26/88.9
Compression ratio	8.3:1	8.3:1	8.9:1	9.0:1	9.0:1	9.0:1
Valve size. in/ex mm	38.1/32.5	38.1/32.5	39.8/34.1	39.8/34.1	41.3/34.1	39.8/34.1
Cam timing	16/56 51/21	16/56 51/21	16/56 51/21	16/56 51/21	20/52 55/17	20/52 55/17
Bhp at 3000rpm*	30	36	38	42	42	44
Maximum bhp*	45	55	60	64 @ 5000	66 @ 5000	64 @ 4800

*at the wheels as observed on a Clayton rolling road dynamometer.

Chapter 2
What is horsepower?

Going back to first principles is the easiest means of linking together, in a logical manner, the various concepts and terms relating to the internal combustion engine and the four stroke cycle.

By far the majority of reciprocating engines use the four stroke or Otto cycle (so called after the originator of the four stroke engine). Each cylinder takes four piston strokes – two revolutions of the crankshaft – to generate one power stroke.

The four cycles are:

1. Induction (or intake). The inlet valve is opened and the piston's movement from Top Dead Centre (TDC) to Bottom Dead Centre (BDC) creates a vacuum that draws the air and fuel mixture from the carburettor, through the inlet manifold and cylinder head port and into the cylinder. To increase the amount of mixture drawn in, the inlet valve is normally opened shortly before the stroke begins and closed shortly after it ends.

2. Compression. Both valves are closed and the piston moving back up the cylinder reduces the volume between the rising piston and the cylinder head that the fresh, drawn in mixture has to occupy. This has the combined effect of compressing the mixture and further mixing the air and fuel together, which allows more of the energy from the fuel to be used.

3. Power (or ignition). The compressed fuel and air mixture is usually ignited by a high voltage spark discharge from the sparkplug prior to the piston reaching TDC (with both valves still closed). The resulting expansion of high temperature and high pressure gases produced by the burning air and fuel mixture then acts upon the top of the piston, which is pushed back down the cylinder, forcing the crankshaft to rotate. This is where the chemical energy in the fuel is converted into mechanical energy. The exhaust valve usually opens prior to the piston

reaching BDC to begin the exhaust process and reduce the pressure in the cylinder.

4. Exhaust. The cylinder is cleared of the burned mixture remaining after the power stroke, initially blown out by the high residual pressure within the cylinder, with most of the remainder being swept out by the motion of the piston back up the cylinder as it approaches TDC. The inlet valve starts to open again as the piston nears TDC, the exhaust valve closes just after TDC and so the process begins again.

The more technical description for the four stroke cycle used by the automotive engineering fraternity is suck, squeeze, bang, blow!

The purpose of internal combustion engines is the conversion to mechanical power of the chemical energy contained in fuel. In the case of spark ignition engines such as the MGB's unit, this takes place inside the engine as the controlled burning of a mixture of air and

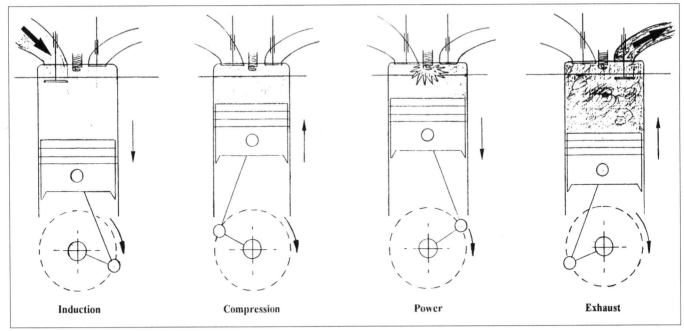

| Induction | Compression | Power | Exhaust |

The four stroke cycle.

fuel that is converted into mechanical power by the engine's internal components. The engine's efficiency in this conversion is known as its performance, and can be defined by the terms torque and horsepower.

Torque is the measure of the turning effort produced on the crankshaft by the pressure from the burning gases acting on the piston during the power stroke. To enable this measurement to be taken the engine is coupled to a dynamometer, a device capable of exerting a force with which to oppose this turning effort (it, in effect, acts like a powerful brake). The amount of opposing force is given as a reading in pounds feet (lb.ft), or the more modern term of Newton metres (Nm). If the engine and the dynamometer are running at the same rotational speed (revolutions per minute, or rpm), then the engine torque must be equal to the dyno's opposing torque. Readings are taken over a range of different engine speeds and then a torque output curve for that particular engine can be plotted.

The term brake horsepower (bhp, or modern term KW or Ps) is the power output of the engine at the flywheel (or output shaft), and is mathematically derived from the torque and rpm readings taken from the dyno.

The work output as measured by a dynamometer also allows the calculation of brake mean effective pressure (bmep). For normally-aspirated engines, bmep shows ability to draw in an air/fuel mixture, how effectively the air is used in combustion, and how completely and efficiently the fuel is converted into energy. In short, it denotes the average pressure acting upon the top of the piston over the complete four cycles of one cylinder. It is a parameter which also includes the mechanical losses that occur within the engine due to friction. As the results, unlike the values for torque and power, are independent of engine size and configuration, they are useful for direct comparison with already established values from other different engine types and designs.

Torque and bmep are directly related, so values plotted for both on graphs give curves with exactly the same shape.

The ability of the engine to utilise the heat output from the fuel is termed Thermal Efficiency. Unfortunately, internal combustion engines are not very thermally efficient and only some 25-30% of the heat generated by burning the mixture is utilised in the form of work. The remainder is lost to atmosphere, mostly out of the exhaust pipe, but also via the coolant and oil or as heat radiation from the engine block.

A portion of the power produced per cycle by the engine is used to draw in the fresh air/fuel charge on the intake stroke, compress it and pump the burnt remains out on the exhaust stroke. Power is also utilised in overcoming the sliding and rotating friction of the internal mechanical components, such as pistons, rings and bearings, as well as to drive the engine accessories such as camshaft, distributor and oil pump.

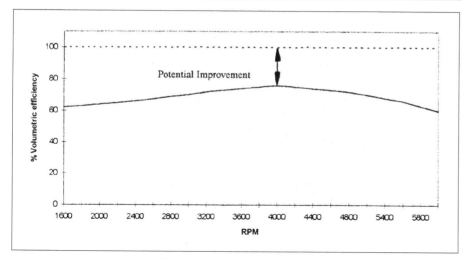

Engine volumetric efficiency.

These losses are all grouped together under the heading friction power. This friction power dissipates useful energy as heat into the oil and coolant.

Overcoming drag from oil whipped up from the sump and acting on the crank and conrods creates another power loss. It, too, falls under the heading of friction, but is also more commonly known as windage loss.

Volumetric efficiency is the most fundamental and important parameter used for defining performance characteristics, as it dictates how much power an engine is capable of producing. It is a measure of the effectiveness of the engine's induction process; its ability to draw the air and fuel mixture into the cylinder. The incoming air has to make its way to the cylinder via the air filter, carburettor, throttle butterfly, inlet manifold, inlet port, and finally past the inlet valve, negotiating bends and obstacles that can amount to a fairly tortuous route. All of this can restrict quite markedly the amount of air that

the engine can draw in. Typical values for modern engines range from 80 to over 90 per cent volumetric efficiency, whilst older designs usually manage around 60 to 70 per cent efficiency. The MG four-cylinder as standard typically falls in the middle 70 per cent efficient category, which leaves plenty of room for improvement!

Careful assembly and attention to detail can minimise the frictional losses mentioned previously, as well as create an increase in bmep which, in turn, will generate more torque and therefore horsepower.

As there is a direct relationship between the mass flow rate of air into an engine and the power capable of being developed, it is vital to get as much air as possible into the engine cylinders. Or, to put it another way, improving the engine's volumetric efficiency is the key to good usable power gains, as the mass of air in the cylinder governs how much fuel can be burnt. Anything that makes it easier for air to get into the

cylinders results in improved volumetric efficiency, generating higher bmep and so more torque and power (provided losses elsewhere have not escalated for some strange reason). In the case of the MGB, a well modified cylinder head (ports and valves reshaped to smooth the passage of air into the cylinder) offers the greatest all-round improvement (to both volumetric and thermal efficiency). Changes to other components, such as the camshaft (which can be thought of as the engine's brain in that it controls the inlet and exhaust processes), to one that opens the valves further and for longer, can also achieve reasonable power gains. But these gains generally occur at a higher rpm than for the standard engine, which, in most cases, means sacrificing power at lower rpm.

The entire engine system really needs to be considered as a whole, otherwise the gains possible from component changes may not be fully realised.

The foregoing should provide a reasonable overview of the various factors used to evaluate an engine's performance and the effectiveness of any changes or modifications made. There are, of course, many more ways of mathematically modelling and evaluating an engine's internal and external processes, most involving complex formulae that are rather beyond the scope of this book. You may have noticed that the maths section has been relegated to the back of this book: all that theory is fine in small doses, but what we're really interested in are the more practical methods of increasing horsepower!

Chapter 3
The engine tuning 'Top Ten' (well, Top Seven, actually!)

Tuning can be an expensive business, and knowing which modifications offer best value on a performance improvement basis can be difficult. To help you, I've compiled a list of what, in our experience, represents a logical approach to MGB engine tuning, though this is purely on the basis of what gives the best performance gains. Most people will not feel a power increase of less than 10 per cent 'through the seat of the pants,' although, oddly, you can recognise a 10 per cent power loss. So the list is based on the biggest performance gains first, and includes items which, by themselves, may not improve performance, but in combination will contribute to the overall performance of the engine.

CYLINDER HEAD

A lead-free conversion has got to be recommended, giving around a 7 per cent power increase with a standard conversion and up to a 30 per cent power gain from a fully modified head (only guaranteed with a Peter Burgess modified head, though!) on an otherwise standard engine in good condition. The real bonus from a modified head is the power gain throughout the entire operating range of the engine. The disappearance of leaded fuel means conversion has become a requirement for these older engines, and not having to purchase bottles of lead replacement fuel additive will help recoup the cost. The engine will obviously need to be in reasonable condition to fully utilise a modified head, and the car will need a rolling road session to set it up correctly – whichever head is chosen.

AIR FILTERS

As this is the first obstacle the outside air encounters on its way into the engine, a change of air filter will make deep breathing a doddle for the engine. Choose from either cotton gauze items from K&N, Green Filters or JR filters, or foam from Pipercross, ITG Megaflow, Ramair or Jamex. Always go for reputable name brands as they filter the air effectively, can be cleaned and re-used and are fire retardant. There may be a slight increase in induction noise, but not unpleasantly so. Carbs will need new needles to correct lean mixture due to increased airflow (see carb chapter) and, ideally, a mixture check on a rolling road.

CAMSHAFT

A change of camshaft offers moderate increases in high rpm power but, unfortunately, mostly at the expense of low rpm performance, so think carefully about your driving needs before overlooking the excellent standard cam. To really get the best results from a cam swap, the cylinder head should be fully modified. Whilst the cam alone may be fairly inexpensive, new cam followers are essential in every case and new valve springs may be necessary to cater for increased lift or engine rpm. The correct installation and timing of the cam is also vital in order to realise any gains, so an adjustable pulley or offset keys will also be needed. Once again, a rolling road session will be necessary to alter the fuelling to suit the engine's specific needs.

EXHAUST

Performance exhaust systems use straight through absorption silencers and smooth curves and bends that provide a less restrictive passage for the exhaust gases, whilst still silencing the noise effectively. They're usually better made and last rather longer than cheap standard replacement brands. Maniflow exhaust manifolds and systems offer improved performance for road and race, or those available from Moss work well for road use. A manifold will have to be included in most upgrades, unless an adapter can be fabricated to enable the very effective standard cast manifold and pipes to be connected.

IGNITION

A change from conventional points to electronic ignition may not achieve any significant power increases, but does do away with the troublesome points that wear out rapidly on a B and can cause poor engine performance due to misfires or erratic ignition timing. There is a wide variety of equally good ignition systems available from Accel, Lumenition, MSD, Piranha, Mallory, Aldon Automotive or Micro Dynamics. This is a performance upgrade that will pay for itself in a short time – no more awkward changing of points for a start! Alternatively, a 123ignition electronic distributor offers a complete solid state ignition in a self-contained package.

A change of sparkplugs to a type better suited for use with any performance upgrade of the engine may also be required.

CARBURATION

A change of carburation from standard SUs to a sidedraught Weber or Dellorto is only really necessary for rally or full race use. If you do want one for road use, a 45 is the size to go for. They offer a very limited power gain for normal road use (1 or 2bhp at high rpm only), being usually less effective than well set-up SUs. The ultimate fuel system would have to be a fully mapped fuel injection system, so you would need to talk to a specialist supplier about such a system if you are interested. Back to carburettors again; a change of inlet manifold will be needed, together with uprated fuel pump and a fuel pressure regulator, not to mention the linkage to operate the carb(s) and, lastly, air filter(s) and ram pipes to suit the application. A rolling road session is also necessary to set the carbs up correctly and check the fuel mixture.

ROLLING ROAD TUNING

Whether a car is modified or not, a tune-up session on a rolling road has to be included in the list of potential improvements. Even if you do all your own maintenance work on the car, a tune-up every year or so by a skilled operator with experience of MGs confirms all is well with engine performance. Very rarely are major problems encountered, it's just a case of checking and fine tuning the engine, which can really put the sparkle back into your car's performance. Prices for a tune-up can vary greatly, so shop around (also check if parts used are extra). If you choose no other form of modification, treating the car to a rolling road tune-up once in a while is guaranteed to put the enjoyment and sparkle back into what may have become a slightly jaded relationship. An example is given later in Chapter 14, in the section titled modern dyno session.

Chapter 4

Power recipes

The performance table for various combinations of components is included to show what can be expected from an MGB engine at different tuning stages.

This should give guidelines toward achieving the performance you want and that will best suit your driving style, motoring needs and pocket!

Power figures are at the wheels, as observed originally on a Clayton Rolling Road dynamometer, and all cylinder heads used are from Peter Burgess.

Engine specification	Power @ 3000rpm	Max power	@ rpm	Comments
Std MGB 1.57 inlet valves	43	65	4800	
Std MGB 1.625 inlet valves	42	65	5000	
Std MGB 1.625 inlet valves, K&N filters	43	68	5000	Cannot feel the difference, but a good starting point for further modifications Plus don't have to buy new filters again!
Std MGB 1.625 inlets, K&N filters & mild camshaft	44	71	5100	Revs well but no grunt
Std race MGB 1.625 inlets. Blueprinted head and engine MGCC, Std. cam 1.5 SUs, K&N filters	60	95	5200	Championship winning engines MGOC, standard race,sprint and hillclimb specification engine by Peter Burgess
1.625 inlets, std lead-free conversion K&N filters (premium unleaded fuel)	47	73	5000	3 angle seats. Feels sharper, plus better mpg
1.57 inlets DIY head, lead-free slight increase in compression ratio (9.5:1), K&N filters	52	76	4800	Torque increase shows as better acceleration. Good for low,sub-4000rpm pulling & cruising
As previous but 1.625 inlet valves.	50	79	5000	Sharper above 4000rpm

Engine specification	Power @ 3000rpm	Max power	@ rpm	Comments
DIY head mods.				
1.625 inlet valves modified lead-free head, 9.75:1 cr, K&N filters.	55	87	5000	Much improved mid-range, good power from 3000-5500rpm
As above + high ratio roller rockers (1.62:1)	59	96	5000	Impressive power, but roller rockers rather expensive addition
1.69 inlet valves, modified lead-free 9.75:1 cr. K&N filters	57	91	5200	Less power than previous below 3000rpm, but very strong above. MPG not quite as good as previously
1.625 inlets, modified lead-free head. Engine bored to + 0.060. K&N filters	58	87	4900	Increased mid-range torque due to bore size. Peak power earlier as port gas velocities reached sooner than with smaller capacity engine
1.69 inlets, modified lead-free head, bore + 0.060. K&N filters	56	90	5100	Feels lively and revs well to 5700
Big bore engine (1.9) with standard 1.625 inlet head K&N filters	50	72	4700	Feels very flat. Needs a cam to peak at standard rpm
Big bore engine (1.9 +) 1.625 modified lead-free with head, K&N filters. Mild road cam	62	95	5000	Getting very quick. Fully modified head opened up chambers
As above with 1.75 SUs	60	98	5000	Increased top end revability! Flatter than previous below 2000rpm
Big bore (1.9 +) 1.69 inlets, modified lead-free head 9.5:1 cr, K&N filters, Peco, Piper. HR270 cam	64	95	5200	Head makes flat 1950 sharper, engine peaks 5200, revs very freely to 6000. Head has chambers opened minimum necessary for flow
Big bore engine, Weber 45 DCOE 1.625 fully modded lead-free head, 715 cam, K&N filter	60	100+	5000	Sounds superb and revs very well. Flat below 2500 and have to ride clutch to pull away smoothly
1.69 inlets, fully modified lead-free, increased cr, Weber 45 carb, 718 cam	58	112	5400	Not very road driveable. First tuned competition level, for sprint and hillclimb use
1.625 inlets, fully modified lead-free. Big bore engine, Weber 45. Piper BP 300 rally cam	55	118	5600	Sprint and rally spec. Won't pull full throttle below 3000rpm. Engine by Gerry Brown, Merton Motorsport
Big bore, big valve, full race engine. Kent 719 scatter-pattern cam (on new rockers, split 48 blank), high lift Webers	won't run	140+	6000	Very quick. Forget it for road use! Peter Hiley's championship-winning race car. Engine by Peter and Andrew Hiley
Big bore (1.9+), fully modified HRG Derrington alloy head 1.625 inlets, 10.5:1 cr, mild cam, SUs	72	95	5200	Very strong mid-range torque. Equivalent to standard 3-litre, 6 cyl MGC power! HRG Derrington alloy head conversion
Big bore (1.9+), fully modified HRG Derrington alloy head. 10.8:1 cr. Fast road cam. Twin 45 DCOE Webers	77	110	5400	Very powerful but very expensive. Extremely rapid!

Rev limits: standard engine 6200rpm; fast road 6500rpm; full race 7500rpm. These guidelines are valid for engines in good condition, or specifically built for an intended purpose.

POWER RECIPES 2 - PROOF OF THE PUDDING!

Since the book was first written, engine components and peripherals plus manufacturing and electronic technologies have advanced and evolved considerably, but the core of the B series remains, and the tuning principles are the same. We too have advanced and evolved, learning a great deal along the way.

We are now able to show graphically the result of component and tuning combinations. While the power outputs given here may not exactly match those in the earlier power recipes section, the original figures do still obtain.

On our new Dynocom Racing inertia rolling road dynamometer the cars were run in direct top, 4th gear, rather than 3rd gear as previous, at wide open throttle and accelerating, rather than being run at full throttle and then loaded down as on the Clayton water brake. All power and torque figures shown are at the wheels and corrected to SAE J1349 – a net rating method as used by new car manufacturers. Where not mentioned otherwise, the horsepower and torque curves are shown in comparison to a standard chrome bumper MGB.

Graph 1 clearly shows the performance improvement (dark lines) above 3000rpm to an engine with an Econotune spec cylinder head fitted with K&N filters when compared to the same engine using standard filters. With the mixture corrected to compensate for the increased airflow the max improvement was 4bhp.

The dyno session to collect data shown in **Graph 2** started from 2000rpm. The engine was not driven through that point, hence the curves starting at zero.

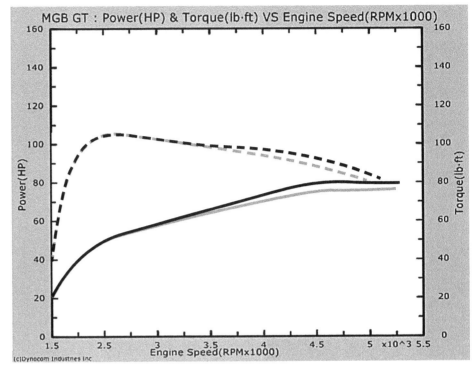

Graph 1: Before and after K&N filters.

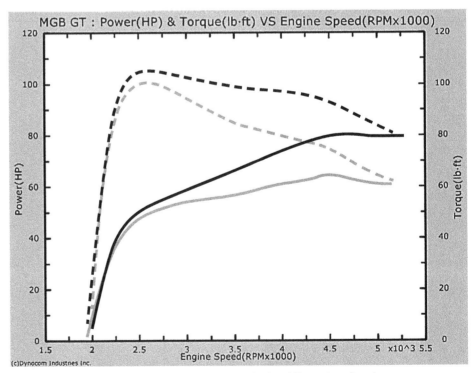

Graph 2: Rubber bumper, standard and Econotune head.

This was an interesting session. It showed the before and after result of fitting of an Econotune head (1.576 inch inlet and 1.343 inch exhaust valves) to a rubber bumper B. More information on this specification of cylinder head is given in the economy section – Chapter 15.

The standard car made very good torque at low rpm. The engine was original and untouched, or so we thought. Removing the head we discovered the cylinder head on this very late rubber bumper B had basic three-angle valve seats from the factory. However, the dimensions of the three seat angles resulted in the inlet throat being rather narrow. The seats would improve low to mid lift airflow, but the narrow throat would restrict overall volume flow and hence top end performance.

We also discovered the head had been removed at some time and skimmed, but the owner was unsure when it could have happened. The car has been in the same family from new and all its service history is known. We have subsequently seen the same on another very late rubber bumper B. We wonder if the work was done during a routine service but not mentioned to customers, possibly as a warranty rectification ...

As predicted, we were hard-pushed to improve the low rpm torque since the compression ratio had already been raised slightly from standard and the factory three angle seats were effective.

With the fitment of the Econotune head torque improved from 99 to 106lb-ft (dark lines). What did surprise us though, was the massive improvement in power and driveability above 2750rpm. Showing 80bhp at the wheels (up from 62 as standard), and with 20bhp as measured losses,

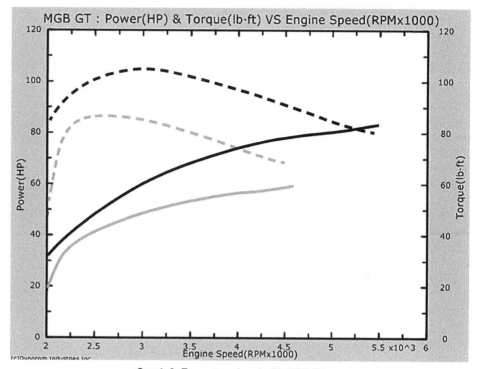

Graph 3: Econotune head with K&N filters.

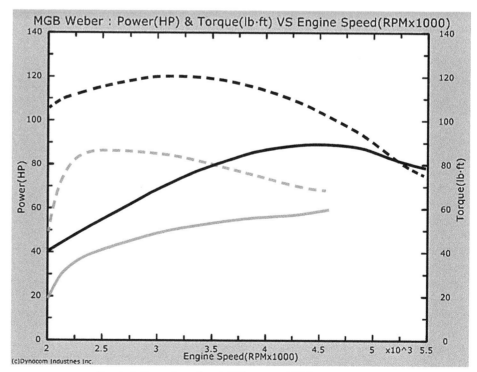

Graph 4: Econotune head, 270 cam and Weber carburettor.

the engine made around 100bhp at the flywheel. Not bad for a head swap!

Graph 3 shows the results of fitting an Econotune specification head, using 1.56 inch inlet and 1.34 inch exhaust valves and a CR of 9.7:1, plus K&N filters on an early chrome bumper MGB. Careful setting of the distributor and valve clearances along with correcting the fuelling upped the max torque from 84lbft to 104lbft, and power rose from 60 to 82bhp at the wheels. Transmission losses were measured at 22bhp, giving a flywheel figure of 104bhp.

Graph 4 is included because of the interesting specification engine, as a sidedraught Weber carburettor is not usually associated with such a 'mild' state of engine tune. This engine is overbored +20, giving 1822cc, and originally the car ran twin 1½ inch SU carburettors with the Econotune (1.56 inlet 1.34 ex) head and Piper HR270 camshaft. The car is used by the customer mainly for cruising on motorways to and from his places of work, so it runs for extended periods at around 70mph. As an experiment, a 45 DCOE Weber carburettor was fitted in place of the SUs. The graph shows the midrange torque from 2500rpm to 4000rpm is very strong when compared to a standard engine (pale lines); ideal for fast cruising and for providing strong pickup without having to change gear. The engine produced a maximum of 90bhp at the wheels, and with measured transmission losses of 20bhp, a flywheel figure of around 110bhp. Of note is that the change from SU carburettors reduced mpg.

Graph 5 shows how much power can be extracted from a standard MGB engine for a standard race class. Rules specify use of a camshaft giving standard lift, though unlimited duration is allowed. It runs

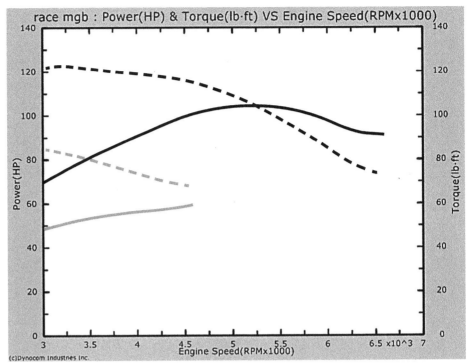

Graph 5: Standard race engine.

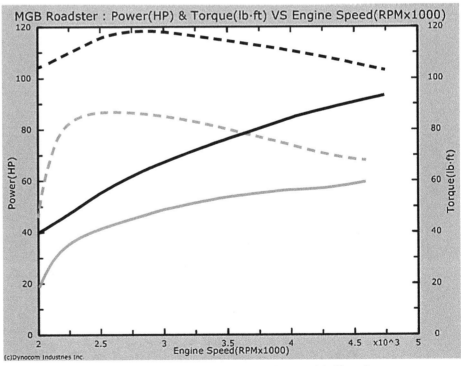

Graph 6: Fast Road head, +60 engine, 270 cam, high lift rockers.

+60 oversize JE Pistons flat top race pistons for 1867cc. The cylinder head is blueprinted to race specification with 1.625 inch inlet and 1.343 inch exhaust valves. Rules also state that while the head casting must show no visible modifications, three-angle seats are allowed. The compression ratio is 11.6:1, and fuelling is from twin 1½ inch SU carburettors. Power and torque can be seen to tail off as the modification limited cylinder head becomes breathless and restrictive. The result of careful engine assembly and attention to detail is 106bhp at the wheels. With 27bhp measured in transmission losses that equates to around 133bhp at the flywheel.

Graph 6 is interesting. This engine is +60, giving 1867cc. It is fitted with a Fast Road head using 1.625 inch inlet and 1.34 inch exhaust valves. It runs 1.625:1 ratio high lift roller rockers (the standard ratio is nominally 1.42:1), a Piper HR270 camshaft and twin 1¾ inch SUs. This MGB engine is actually fitted in an MGA.

The torque curve is very flat, peaking at 118lb-ft at 2800rpm. Peak power is 96 at the wheels at 5000rpm. The transmission losses were 22bhp, meaning overall around 118bhp at the flywheel. A lovely torquey package that is very useable on the road or for track day fun.

The data in **Graph 7** was obtained from a 1867cc (+60) engine fitted with a Fast Road head with 1.625 inch inlet and 1.343 inch exhaust valves and 10:1 compression ratio, HR285 cam and electronic ignition. It runs twin 1½ inch SU carburettors with K&N filters. Between 3000 and 4000rpm it shows the characteristic torque dip of an HR285 cam and SUs. It would make 6 or 7 more bhp fitted with twin 1¾ inch SUs. With measured transmission losses of 26bhp and 94bhp at the

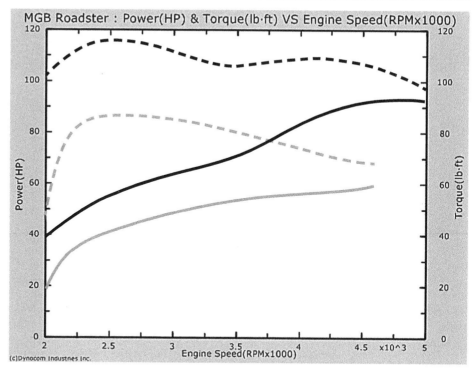

Graph 7: Fast Road head, +60 engine, 285 cam.

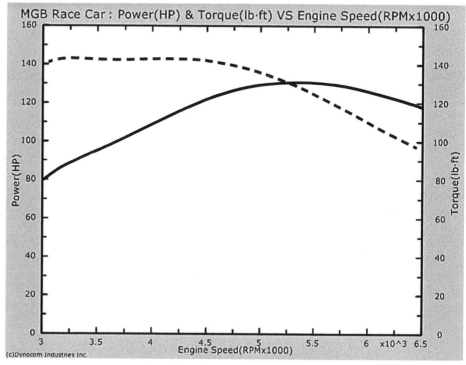

Graph 8: 1.625 inch inlet race engine.

21

wheels, this B made around 120bhp at the flywheel. A very willing and revvy engine.

The engine used to produce **Graph 8** runs +60 race pistons (1867cc), and a race head using 1.625 inch inlet and 1.343 inch exhaust valves. It runs a compression ratio of 11.6:1, a full race cam, and utilises a crankshaft oil scraper to further enhance output. While it is class restricted to twin 1½ inch SU carburettors, it still makes really good power. Above 5500rpm the carburettors are past their best in terms of airflow capacity and so bhp/torque falls fairly rapidly. With a measured 24bhp in transmission losses this engine produced around 154bhp at the flywheel – remarkable given the valve sizes used.

The 1840cc engine shown in **Graph 9** started life in supercharged guise. It was built with a Piper BBP270 camshaft, race pistons and a supercharged (SC) specification cylinder head (1.625 inch inlets and 1.343 inch exhausts). When the supercharger gave up the ghost the owner reverted to naturally-aspirated. He skimmed the head to get a compression ratio of over 10.5:1, fitted high lift 1.625:1 ratio roller rockers and a Weber 45DCOE with 40mm chokes. The resultant power and torque output was very respectable. The best run saw 115bhp at the wheels. The transmission losses were 20bhp, giving around 135bhp at the engine. A lovely track day performer.

The dark lines on **Graph 10** show the improvement in bhp and torque compared to a standard MGB (pale lines). The more powerful engine has a 1950cc bottom end fitted with a Fast Road specification head using 1.625 inch inlet valves and 1.343 inch exhaust valves, with a compression

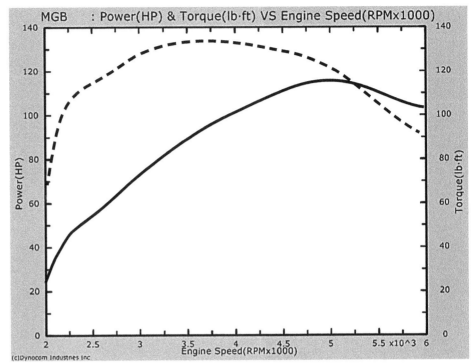

Graph 9: Ex-supercharged 1840.

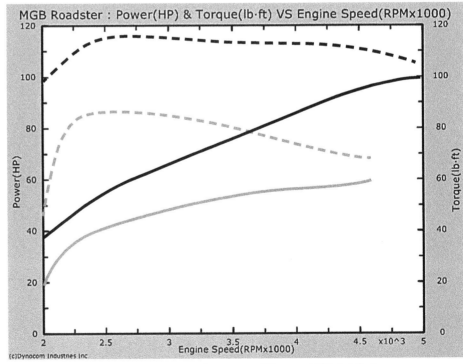

Graph 10: 1950cc, Fast Road head, 270 cam, 123T distributor.

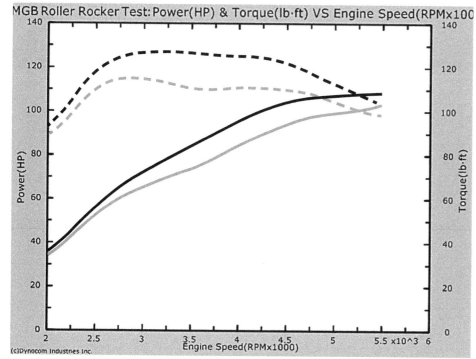

Graph 11: 1950cc, Fast Road head, 270 cam, high ratio rockers, 123T distributor.

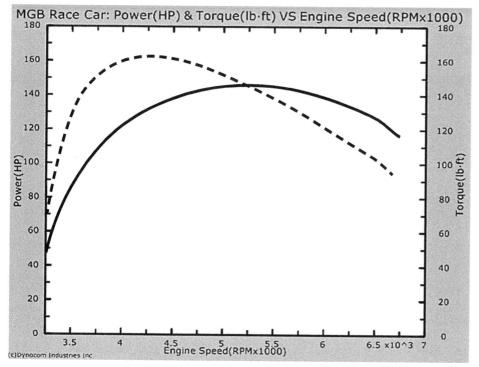

Graph 12: Endurance 1950, Weber carburettor.

ratio of just under 10:1 and a Piper HR270 camshaft. It is fitted with an electronic 123Tune distributor and runs twin 1½ inch SU carburettors with K&N filters. Note the particularly long, flat torque curve, meaning it will pull like a steam train. The engine peaked at 104bhp at 5500 rpm. With a measured 26bhp in transmission losses, this meant around 130bhp at the flywheel.

Graph 11 shows the same 1950cc engine fitted with 1.625 ratio roller rockers. When setting up this combination we found the mixture had changed from the previous dyno tests, having become richer. This was due, we think, to the stronger intake pull generated by the quicker valve lift. Weakening the mix off slightly, back to the pre-rockers readings, proved beneficial.

Connecting a laptop to the 123Tune distributor made optimising the ignition quite straightforward. It resulted in us retarding the timing two to three degrees across the range to find the sweet spots, the engine not requiring as much ignition advance as previously. We feel this was due to the increased cylinder filling (volumetric efficiency) delivered by the rockers and cam combination.

The accompanying graph shows the at the wheels results before (pale lines) and after (dark lines) fitment of the roller rockers. The engine gained power throughout, but the real eye opener is the mid to upper range torque increase.

Graph 12 shows an Endurance racing 1950 which delivers a broad spread of very useable power. The class limits engines to standard (1.42:1) rockers and a single Weber carburettor (split Webers are not permitted). The camshaft is a Full Race grind and the compression ratio is 11.8:1. The inlet valves are 1.78 inch,

23

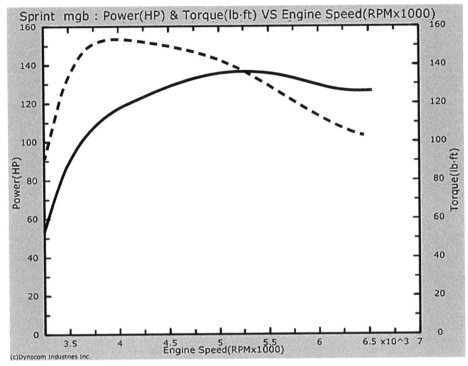

Graph 13: Sprint 1950, Weber carburettor.

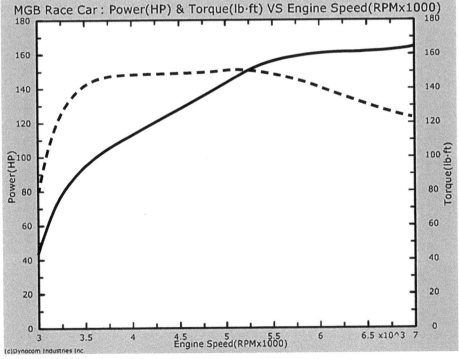

Graph 14: 1950 with Split Webers.

the exhaust valves are 1.343 inch, and the inlet ports are as large as we dare make them without ruining the head by breaking into waterways or making it unreliable. The engine runs conventional contact breaker points distributor ignition as specified for the class. Giving 144bhp at the wheels and with transmission losses of 29bhp, this B engine has around 173bhp at the flywheel.

The 1950 engine from **Graph 13** was built for sprints and hill-climbs where torque is vital for exiting corners and dealing with the gradients! It uses the old Leyland Special Tuning 770 profile camshaft – milder than full race, yet one which always works well with this type of engine. The sprint specification head uses 1.69 inch inlet and 1.343 inch exhaust valves. The CR is 11.4:1. In wet conditions this specification of engine can win at short circuit racing. It runs a 45 DCOE Weber with 40mm chokes.

Graph 14 was produced using the best 1950 race engine we have seen to date. It runs split 48 DCOE Weber sidedraught carburettors – only one choke of each carb is used, aligned to feed straight into each inlet port – and a full race cam and head with 1.75 inch (44.5mm) inlet and 1.343 inch (34mm) exhaust valves, and 11.8:1 compression ratio. The lower rpm performance is not as good as the endurance 1950 but the high rpm power and torque is very strong. This is ideal for long track circuit racing. 163bhp at the wheels and 25bhp measured power losses gives 188bhp at the engine.

Graph 15 shows the effortless power output of a supercharged 1950cc engine. Running 10psi boost, a supercharged specification head using 1.625 inch inlet valves and 1.44 inch exhaust valves, 8:1 compression

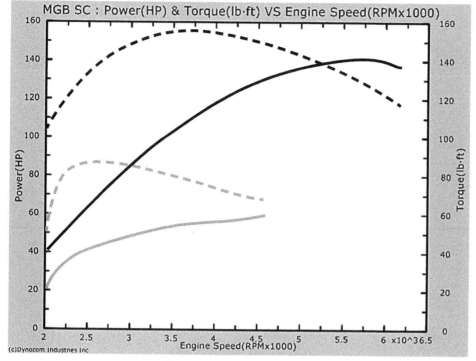

Graph 15: Supercharged 1950.

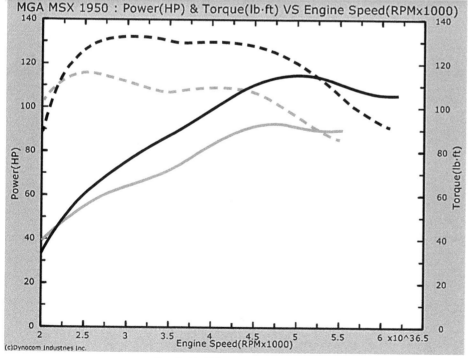

Graph 16: MSX 7 port head 1950.

ratio and a supercharger spec fast road camshaft. Showing 141bhp at the wheels with 25bhp in transmission losses, this engine produced around 166bhp at the flywheel.

The alloy MSX 7 port X-flow head 1950 in **Graph 16** works well when modified, giving very good mid range power and performing strongly to 6000 plus rpm. The engine featured in the graph (dark curve) is fitted to an MGA. It is 1950cc with a Fast Road specification 7 port head using 1.62 inch inlet and 1.34 inch exhaust valves and having a 10.25:1 compression ratio. It is fitted with a Piper HR285 cam and a pair of Weber 45DCOE carburettors.

For comparative purposes, the pale power curve on the graph is an 1870cc engine with a Fast Road specification cast iron 5 port head with 1.62 inch inlet and 1.34 inch exhaust valves, Piper HR285 cam, and twin 1½ inch SU carburettors.

The MSX head produced slightly less power than the cast iron headed engine below 2000rpm due to the size of carburettor chokes fitted. With smaller chokes, more low rpm power would be obtained, though at the expense of high rpm output. The MSX 1950 made 115bhp at the wheels, which with 20bhp transmission losses, meant around 135bhp at the engine. The 1870cc with FR head made 93bhp at the wheels and 18bhp losses, giving about 111bhp at the engine.

Having a rolling road facility offering high speed data acquisition has become very important in enabling us to ensure engines are set up to run at their best.

The power and torque differences from a fuelling change are clearly visible on **Graph 17**, which includes the air fuel ratio (A/F) trace on the lower half.

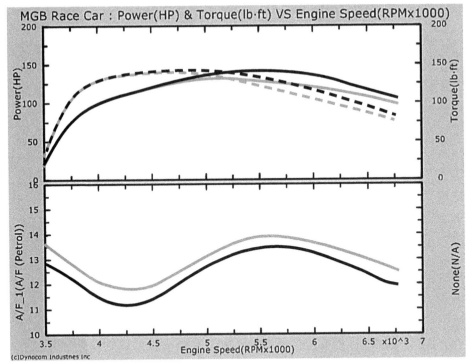

MGB Race Car : Power(HP) & Torque(lb·ft) VS Engine Speed(RPMx1000)

(c)Dynocom Industries Inc.

Graph 17: Effects of fuelling on a race B.

The dark line shows a richened fuel mixture and correspondingly more high rpm power. Yet here we are at a point of losing power between 4000 and 4500rpm where the mixture was at its richest. This race engine was running a Weber 48 DCOE carburettor fitted with 42mm chokes that was struggling to supply sufficient fuel at high rpm, even when fitted with minute air correctors. We have noticed some of the newer Webers do not calibrate as well as some of the older ones!

Chapter 5
Cylinder head

The cylinder head can be considered the key to success or failure of an engine design. If an engine is equipped with a poor cylinder head design in terms of airflow or combustion chamber shape, then its ability to produce power is immediately limited. The four-cylinder B series has been blessed/cursed with a 5 port design that uses a pair of siamesed ports to supply the inlet valves, leading into a good, effective combustion chamber shape. The centre pair of exhaust valves share a common port, with only the exhaust valves at each end of the head having individual ports. When it comes to producing power, it's my opinion that the cylinder head is the real Achilles heel of the B series engine.

There have been numerous changes to the design of the head castings throughout the engine's lifespan, though all have used either 1.56 inch (39mm) or 1.625 inch (41mm) inlet valves and 1.34 inch (34mm) exhausts. These detail changes have affected mainly the port and combustion chamber shapes. As it stands, the head is quite an effective design, despite all the bad press its received, only being let down by poor casting quality – and in some cases core shift causing slight port misalignment.

Don't worry if you don't have one of the standard head castings considered by some to have a high performance

AIRFLOW (CFM at 25in H2O)			
	A	B	C
Valve lift (thou)	1326 casting 1.56 inlet valve	4736 casting 1.56 inlet valve	2709 casting 1.625 inlet valve
50	20.6	19.6	18.4
100	39.8	39.8	37.9
150	58.5	57.8	56.9
200	73.8	74.6	73.3
250	86.5	87.5	88.2
300	96.2	97.5	99.8
350**	101.2	105.1	108.9
400	103.8	109.3	114.3
450	105.1	111.4	117.6
500	106.3	112.7	119.5

**Standard cam lifts to around this point

Table 2: Airflow comparisons for standard MGB cylinder heads.

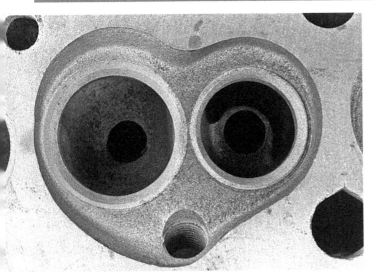

The more open combustion chamber of a 2709 casting ...

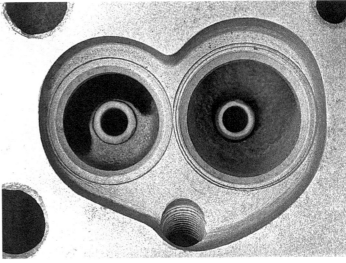

... compared to the beaked chamber of the likes of the 1326 casting.

reputation. The only important physical differences of valve size and combustion chamber volume that exist between the various castings can all be negated by the modifications described.

Columns A and B in Table 2 on page 27 show the airflow for a standard inlet port (1326 and 4736 heads) fitted with 1.56 inch (39mm) valves; not too bad for standard. Column C shows the flow for the later head (2709) with the 1.625 inch (41mm) valve, again, tested as standard. In all cases the valve seats are too wide to give good airflow at valve lifts above 100 thou. The smaller valves flow slightly better up to 200 thou lift, before losing out to the larger valve at higher lifts. The ports and chambers are not well cast, with lumps and bumps everywhere, giving rise to uneven airflow between the cylinders which causes the characteristic lumpy tickover on a standard B.

It also needs to be said that the only

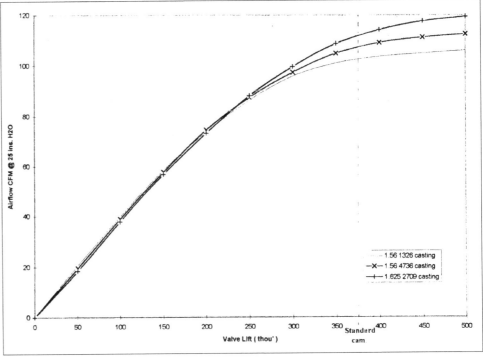

Airflow comparison of standard B series cylinder heads.

similarities between the B series cylinder heads and those fitted to the smaller A series engines are they are both of a five port design, are made from cast iron and are both used on internal combustion engines – nothing else! They are wholly dissimilar and should be treated as such. When it comes to modification, it should not be assumed that whatever works well on the A series heads will also produce the same results when applied to the larger B heads.

AIRFLOW REQUIREMENTS

Successful cylinder head modification first needs to be approached from the point of view of airflow. In order to determine what does or does not work in terms of airflow, all the parts involved in getting air into and out of the engine need to be tested on a flowbench: a piece of equipment that allows accurate and repeatable measurement of the volume of airflow through a component.

Refinement of the flow passages and component shapes involves first establishing a baseline by testing the standard item in unmodified form. In the case of the cylinder head, the port flow is measured at regular increments of valve lift. It's usual to use every 0.050 inch from zero to 0.5 inch (or more if the valve lift of a particular engine is greater). This is followed by a series of gradual modifications and changes to the head, each evaluated for effect on flow and used as a guide to further modification. This is often a long process, involving a great deal of painstaking work: consistently reproducing the end results of all this testing across all the ports in a head, and for many heads, needs a great deal of skill and experience (not to mention professional equipment).

Fortunately, you can bypass some of that work and succeed in improving airflow – with a little effort and patience – by following the guidelines supplied for straightforward and effective cylinder head modification, the effects of which are noted in the accompanying tables.

The fundamentals of airflow (which, in fact, apply throughout the entire inlet and exhaust system) may, at first, seem somewhat strange, but to know them will help in understanding cylinder head modification.

Airflow is more sensitive to shape than size, so big ports are not necessarily better at flowing air than small ports. Airflow also dislikes sudden changes in direction, volume and/or shape. These concepts are supported by the fact that the areas of the port which are easily accessible normally have a moderate effect on airflow, and it is more often those really difficult-to-get-at bits that usually have the greatest influence on the head's airflow capability. Being able to modify those difficult bits in an effective manner is where the art of cylinder head modification comes in.

SAFETY FIRST!

Before we get down to the nitty gritty of how to modify heads, a quick word about safety is required. **Always** wear eye protection when grinding heads. Safety glasses are the absolute minimum requirement, and even then the dust and grit whizzing about while grinding can still find its way into your eyes. Goggles or a full face shield are better. Protective handwear is advisable. Wearing a facemask is also advisable, as inhaling the dust is not to be recommended. Ear defenders are optional; it depends on how noisy your equipment is, but they are generally a good idea.

GRINDING EQUIPMENT

Having equipped yourself with safety gear you also need to find somewhere suitable to work, preferably well lit (an old skeleton desk lamp makes an excellent movable light source for illuminating awkward to see parts of the head). Grinding dust also makes a dreadful mess and can spread over quite an area (especially when using front exhaust air grinders), so be warned!

Next, you need something to grind with. The modifications described here can all be done using an electric drill and grinding stones (also called mounted points). To be fair, the drill is not intended for this type of use as it's too slow for easy removal of material; the large chuck obscures visibility and limits port access. Also, the drill's bearings are not intended to take the side loads caused by grinding and can wear out rapidly if used in too

Modifying an MGB head.

brutal a manner. But a drill is perfectly adequate if you only intend to modify one or two heads; it just requires a little care and patience to get results.

The alternative is to splash out on some professional equipment, in which case you can choose either a hand-held electric die grinder, a remote mounting flexible shaft electric grinder or, if you have access to a compressed air source, a pneumatic die grinder. This equipment is not cheap to purchase though and, in some cases, the prices can be close to those of a professionally modified head. If you are going to modify more than one head yourself, though, it becomes a worthwhile investment.

The best mounted points to get for grinding cast iron are Aluminium Oxide – 60 grit stones (the pink ones). These come in a wide variety of shapes and sizes. You will need a few of the smaller oval and round stones to begin with, but the large, round, one inch balls represent better value. Once they wear down you're left with a collection of different shapes anyway. Some means of dressing the stones is also advisable to allow minor reshaping and to enable the stones to be balanced before use. Every time they are fitted to the chuck the stones must be dressed until the worst of the vibration goes away, otherwise they 'chatter' in the port and can do more damage than good. For smoothing and squaring the rectangular exhaust ports, stones of about 0.25 inch (6mm) diameter by 1 inch (25mm) long will allow you to get into the corners more easily. A couple of 1 inch diameter parallel stones complete the collection. These have flat faces and parallel sides and are ideal for tidying up the combustion chambers. All of these stones are available from trade suppliers (**Caution!** The stones from DIY stores are often not safe at high speeds), together with carbide cutters which, again, are available in a variety of

A selection of aluminium oxide stones and carbide burrs.

shapes but cost considerably more than the mounted stones. Carbides remove metal quite rapidly (though, to be really effective, they do need higher speeds than a drill provides), and do not change shape with use so curves can be easily duplicated. They also do not wear out like stones, but do gradually become blunted with use. Even so, they are a worthwhile purchase if you intend to modify several heads.

In order to achieve a smooth finish after grinding, use a rod approximately 2.36in (60mm) long by 0.24in (6mm) diameter silver steel with a 1in (25 mm) slot cut lengthways)with a short length of 25 mm, wide 60 grit aluminium oxide cloth (or emery cloth) wrapped around it. This works wonders for removing any minor imperfections in the porting work and leaves the perfect matt surface finish.

When porting, keep the grinding stone moving at all times. Leaving it too long in any one place will create a hollow that will have to be removed: if you're not careful you could end up chasing hollows all around the port and finish up with a mess. Just a moderate pressure on the stone is all that's needed, and keep it moving. Finishing the port off with a thick roll of the oxide paper on the split rod will remove any remaining ripples after grinding. A good matt finish

is all that's required. A highly polished port does absolutely nothing to improve power, so why waste time trying to achieve it?

With that little lot out of the way, we can get on with the job of working on the head, although we're by no means ready to begin porting yet!

INSPECTION AND CLEANING

With the head off the engine, and after it's stripped, you can perform the first visual inspection to see if all is well and the head is actually worth working on, or if it's necessary to find another. You can also carry out these checks if the head is only off the engine for a de-coke, and so nip any potential problems in the bud.

The studs are removed from the head and manifold faces using a stud extractor, or a low-tech alternative is to lock two nuts together on the stud and unscrew with the lower nut. Any studs that snap off (the thermostat housing studs are the more frequent offenders) will have to be drilled out and have a Helicoil thread insert fitted, unless you're a dab hand with stud extractors. Fitting Helicoils is really a job best left to the professionals.

All the B heads can crack at the front face of the head, between cylinders two and three. This shows up as a rusty streak or a line of white powdery deposit where anti-freeze has seeped through, flaking off the paint on the head. The crack usually runs between the sparkplug hole and the stud hole at the top. When they really go, the crack goes right round from the valve guide in the chamber to the same valve guide at the top of the head. In either case the head is scrap. If a crack is suspected, it is best to get it checked with a dye penetrant, available from engineering trade suppliers or try your local engine reconditioner. With the 'lump' head it

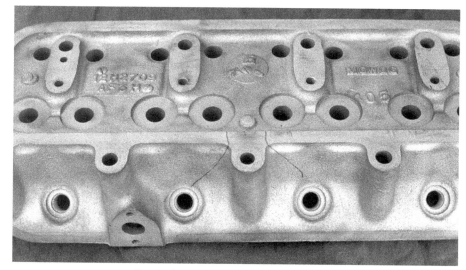

Cracks in the sparkplug side of the head.

any carbon from this region in order to see – coarse emery paper is ideal for this purpose. These lesser cracks can be repaired using replacement valve seat inserts, but you will have to go to a head specialist or engine remanufacturer to get the work done.

In the case of the smaller inlet valved 1326 casting, cracks can also travel from the exhaust valve seats (of number two and three cylinders in particular) out across the combustion chamber towards the chamber wall. These can also be repaired by fitting exhaust seat inserts, provided the crack is not too bad. If it has actually reached the wall of the combustion chamber then

Most often a suspected crack such as this one across the gasket rail and toward the plug hole only becomes apparent after careful polishing.

The beads of water visible between the valve springs gave the game away in this case.

is often surmised, incorrectly, that the casting has been thickened here in order to stop this cracking problem. This was, unfortunately, not the designers' intention. These heads are only an ordinary version of the air injection heads that were to go on engines intended for the American market and used on domestic engines due to rationalisation of casting manufacture.

These heads still crack in the same location, from one side of the 'lump' (or air injection boss) across toward the sparkplug hole, and are still scrap as a result.

Another common place for cracking to occur is in the combustion chamber. Check the narrow bridge of metal between the inlet and exhaust valve in all the chambers. You will need to remove

it's not a viable repair job and the head is scrap.

Having passed the mass of initial checks, the head needs to be thoroughly cleaned. Here, access to a chemical wash tank is useful, as the caustic solution in a chemical tank not only removes caked-on oil, gaskets and any loose paint, it also cleans out years of corrosion and accumulated detritus

Cracks in combustion chambers and valve seats.

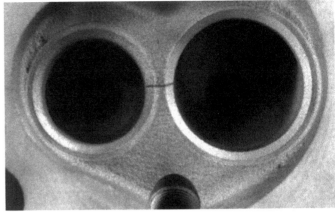

Crack between valve seats.

from the waterways in the casting. The corrosion acts like an insulating layer, reducing heat transfer to the coolant and creating hotspots where coolant flow is restricted, which may bring about the cracking problem with the castings. Removal of this corrosion has a very beneficial effect on cooling around the ports and chambers, as it allows the water to circulate more freely. The best method of removing the corrosion is to blank off the thermostat and heater take-off holes and fill the waterways with a corrosion remover and de-scaler of the sort used to clean central heating systems. Leave it to soak for a good while and then thoroughly flush out the head.

An old washing-up bowl and brush, with a cleaner such as Gunk, works well at removing all the old oil and bits of gasket from the head, if a dip in a caustic tank is not possible. When it's thoroughly dry afterward, the head needs to be grit blasted in order to remove all the remaining baked-on carbon from the ports and chambers. A wire brush mounted in a drill will also dislodge much of the carbon if blasting is not possible. Once clean, the head is ready for a further visual inspection, just to ensure nothing was missed with the first checks.

VALVES

Next, establish whether the valves are worn and in need of replacement. First, check all the valve stems for wear and damage such as ridges or scoring by lightly running finger and thumb along the length of the valve stem, working all the way around it. Any slight ridges should be readily apparent (you can feel wear of a thou or less). If any wear at all can be felt, then new valves are in order. The alternative method is to measure the valve stems with a micrometer. Check in several places around the stem, concentrating mainly in the region that actually runs in the guide. Compare this with the measurements in the workshop manual, and, if the valves are at all worn you need to get new ones.

Now would also be a good time to consider the level of modification that you'd like to undertake with the head, keeping in mind any other intended changes to the engine.

If the head has the 1.56 inch (39mm) inlet valves as standard, if you plan to retain the standard cam profile in the engine (which is a very effective grind – see Chapter 7), and if you don't use the engine much over 5000rpm or so, then there's no need to change the valve size. Retaining this valve size will give the engine improved flexibility and

performance right where you need it, through more low- and mid-range power.

On the other hand, if the head already has the 1.625 inch (41.28mm) inlets fitted, or if you wish to improve middle to high rpm power from your engine, as for use with a performance camshaft (although this applies equally to the standard cam profile, too), these larger valves are the ones to go for.

The inlet valves fitted as standard are made from EN52 steel; a material with good wear resistance, not requiring chrome plating on the stems or hard valve tip inserts for where the rocker pad contacts the valve.

Going larger still with the inlet valve sizes is really only necessary for rally or race applications, or to give large bore road engines (1.9 +) real performance and sparkle (they can seem rather flat otherwise), or if you wish the ultimate for road use with a smaller engine – where they are combined with performance cams and further engine uprating. It means a great deal more modification work is necessary on the head to allow these valves to achieve anywhere near their flow potential. As a result, the smaller bore engines will lose some low rpm tractability, more noticeable than with the bigger bore engines, but given as a trade-off for more top end

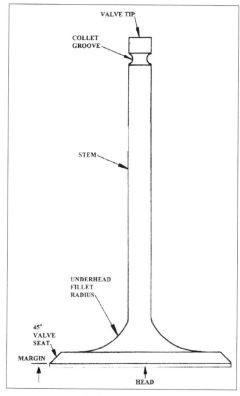

Valve anatomy.

From left to right: modified standard valve, custom big valve, standard valve.

power. It also means more expense when compared to purchasing standard replacement valves, as the larger valves will have been custom made, from either EN52 steel or 214N stainless, and that doesn't come cheap ...

While we're on the subject of valves, it would be a good time to cover the selection of exhaust valves available. The specification of material used for the standard exhaust valves in B series heads is 214N Austenitic stainless steel. This is the best specification of material for the job, well capable of withstanding the rigours of unleaded fuel. Note that some reconditioned heads may have had new exhaust valves of a lesser material specification fitted. If the existing valves from the head are not too worn and you are intending to re-use them, it's possible to check the material by holding a magnet against the valves. The Austenitic

stainless steel used for valves is non-magnetic, so if the magnet sticks the valves will need changing.

As with the inlet valves, the exhaust valve stems are chrome-plated and come with hard-wearing Stellite tips as standard.

For normally aspirated applications we've found it unnecessary to fit exhaust valves of a larger diameter than the standard 1.34 inch (34mm), even in championship-winning full race heads. So if standard exhaust valve size works well in the case of race engines, which operate at rpm levels far higher than road engines will ever be taken to, there's no real need to go any larger.

That said, larger exhaust valves do prove beneficial for forced induction applications. We typically use a 1.44 inch (36mm) exhaust valve, in combination with a 1.625 inch (41.28mm) inlet to provide additional exhaust flow capacity, the intent being to remove any possible bottleneck for the expulsion of the additional volumes of exhaust gases present under boost.

Don't throw the old valves away, though; they can be used as masks for the seats when modifying combustion chambers.

VALVE GUIDES

With valve guides there is a choice between standard cast iron or bronze alloy replacements. The standard guides are fine for use with a standard head, but as we're interested in improving the head's performance the bronze guides are definitely the ones to go for. This is especially true if unleaded fuel is going to be used.

Bronze guides offer a far greater ability to transfer heat away from the valve stem, have better wear characteristics, and can run with reduced lubrication, when compared to the cast iron variety. Bronze guides also run tighter valve clearances, so less oil gets down the guide to contaminate the incoming mixture or carbon-up the back of the valve. After many years of research and development we now use special manganese silicon bronze alloy valve guides in all heads, made from a material specifically developed for valve guides. These guides are suitable for running either chrome stem or non-chrome stem valves equally well, unlike some types of bronze guides that can gall and wear badly if chrome stem valves are not used. Always ask a supplier what valves should be run with the bronze guides they are selling to avoid any wear problems later. If in doubt, get valves with chrome stems – inlets and exhausts.

It is their ability to transfer heat away that is the most important aspect of using these bronze guides, with particular consideration given to running unleaded petrol. With the reduction in amounts of lead in fuel (the UK once had 0.4g lead/litre – from 1986 it became 0.15g/litre leaded, 0.013g/litre unleaded – and there will be less in the future, no doubt), the petrol companies are increasingly turning to other additives to try and duplicate its effect. In case you did not know, lead – or rather, tetraethyl lead – as well as lubricating the valve and seat in the head, also acts as a flame retardant and slows down the rate (speed) of burning of the mixture in the cylinder. Without the lead, the fuel burns much faster and hotter (hence the need to retard the ignition timing for cheaper, lead-free fuel – not super unleaded, though – to compensate for this), giving the valves a hard time. The exhaust valve has to endure by far the worst conditions with extreme heat from the burnt gas passing around the valve on its way out of the engine, as well as continual pounding against its seat in the head. The inlet has it somewhat easier, being cooled by the lower temperature, fresh mixture as it is drawn into the cylinder.

Bronze guides do cost more than the standard cast iron replacement variety, but there is no point in skimping

Inlet valve guides. Left to right: cast iron, bulleted bronze, standard bronze.

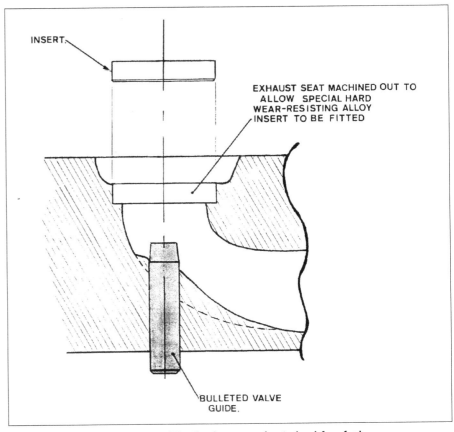

INSERT

EXHAUST SEAT MACHINED OUT TO ALLOW SPECIAL HARD WEAR-RESISTING ALLOY INSERT TO BE FITTED

BULLETED VALVE GUIDE.

Valve seat modification in conversion to lead-free fuel.

when it comes to modifying your head: fitting them is definitely recommended. They come initially in the form of straight bronze tubes, which are fitted by being pressed into the head and then reamed to give the correct valve clearance. Reaming after fitting is essential to remove any distortion from the guides that has occurred during fitting, and to guarantee correct running clearance.

When the guides are for use in a modified head, you may wish to get them bulleted before fitting, which helps to reduce the obstruction caused by the guides where they protrude into the port. Four of the guides should be machined with a 7 degree taper from one end, for a length of 0.43in (11mm). These will be the inlet guides. It is only necessary to taper the exhaust guides if grinding the cast guide bosses down during porting. Then the remaining four guides should have the same 7 degree taper machined for a length of 0.27.5in (7mm), ready to take exhaust valves.

If you are going to modify the head it is by far the best idea to carry out some of the grinding work in the ports and on the guide bosses before the guides are fitted and the seats are cut. This eliminates any chance of accidentally damaging the new seats and working in the ports is a little easier without this worry. Otherwise hitting a seat with a carbide cutter or a stone will mean the seat will have to be re-cut, which may cause problems later when balancing chamber volumes, not to mention cost more money.

VALVE SEATS AND LEAD-FREE FUEL CONVERSIONS

Whether the head was found to be cracked around the valve seats or not, I would strongly recommend having hardened inserts fitted in place of the standard exhaust valve seats. The age

of most MGB cylinder heads, as well as the hard life they lead, usually means that most heads will already have been reconditioned and have had the valve seats re-cut at some stage. Those heads that have not been reconditioned will be in generally poor condition due to the exhaust valve seat erosion and damage caused by today's low-lead fuel. Cutting new seats in the head will cause them to sink further into the chamber, which means the valve sits higher in the head and the valve train geometry can be upset. These inserts will act as protection against increased wear, running equally happily on leaded or unleaded fuel – though you may as well run the less expensive unleaded fuels. That way you save money and the conversion will help pay for itself!

On the subject of valve seats, the only ones to have on any head, performance or otherwise, are what are called three angle seats. These comprise three angles, as the name suggests, a 60 degree bottom cut that blends the valve throat into the 45 degree valve seat, followed by a 30 degree top cut that merges the valve seat into the

Right: Machining recesses for valve seat inserts.

Below: Valve seat insert ready for fitting.

chamber. It is this region of the cylinder head where the greatest gains in airflow can be realised. As mentioned before, airflow hates abrupt changes in direction or volume, and the path from the port past the valve causes the greatest obstruction to flow in a head. So anything that can be done to smooth the path of the air from port to chamber has got to be worthwhile. Compared to conventional (45 degree) single angle seats, the benefit of three angles can be an immediate 25 per cent increase in flow! It takes some mighty fancy porting work with the single angle seat to match these flow gains, more often than not the fancy porting fails miserably in the attempt (actual verification of the increase on a flowbench would be necessary) – and you pay for the time taken to do it, too! In some cases – certain classes of racing, for instance – that porting work is not allowed.

The new seats in the head should be cut so that the valve seats around the outer edge of the valve seat and not the middle or inside edge. This increases the valve seat diameter and therefore maximises the width of the port throat in the head by producing a wide 60 degree bottom cut. When modifying the head this broad 60 degree bottom cut can be reduced in width by grinding out the throat with a stone, so increasing the diameter of the throat and allowing the creation of a smoother shape for the airflow to follow. The bottom cut serves as an excellent visual aid for grinding work, acting as guide to enable the throat to be opened out evenly all around, and providing a nice visible lead-in before getting close to the seat itself. Occasionally an additional 75 degree bottom cut is employed when using much larger inlet valves and inlet seats in alloy heads.

There are a variety of different seat cutter widths available, the most common giving a 1.5mm seat, a 1.3mm seat or a 1mm seat – intended for race use. The

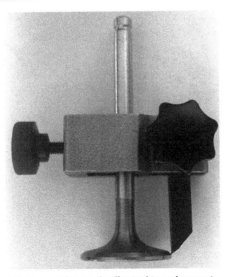

Cutter setting tool adjusted to valve seat diameter.

Valve seat cutter adjusted to setting tool to give correct seat diameter.

majority of road use seats are 1.3mm wide, a big difference when compared to the 5mm wide seats of the standard head. The 1.3mm seat width used is sufficient to allow the essential heat transfer to take place between the valve and the cylinder head when the valve is closed, as well as to ensure the mechanical reliability and longevity of the seat, yet allow for improved airflow through a wider throat. That said, flow through very large inlet valves can be further enhanced using 1.5mm or 1.7mm 45 degree seat widths.

Cutting 3-angle valve seat with Mira seat cutter.

Selection of 3-angle seat cutters.

The seats should also be cut in such a way that all the valve heads finish up at the same height, within a tolerance of plus or minus two thou. If the valve heads sit at differing heights in the chambers it will have a marked effect on capacity and so make life difficult later when it comes to equalising the chamber volumes in order to balance the compression ratios of the individual cylinders.

BLUEPRINTING AND BURETTING

Now we can finally get down to describing the modifications that can be done to the head.

To begin with we'll cover the most apparently straightforward cylinder head for a B, the blueprinted standard race head. This specification of head is used on engines of cars competing in the MGOC and MGCC standard race and hillclimb classes. The regulations for these classes state that absolutely no modification to the ports or valves is permitted, so no porting or polishing is allowed whatsoever, but valve seat specification is free and conversion to run unleaded fuel is allowed. Having the lead-free conversion done allows use of higher octane super unleaded petrol, which means the engine will take more ignition advance and produce more power.

The no modifications regulation means that a specially selected casting is necessary in order to comply with these rules and yet still allow the engine to produce race-winning power. An appropriate head is usually one from the later B series 2709 castings, which come with the larger inlet valves as standard. The head also needs to have the absolute minimum of casting faults and/or core shift visible, so the throats of the ports are not misaligned with the valves and do not have rough, misshapen walls. It is also vital that the valve seats are not badly damaged or sunken in the chambers, as cutting the new valve seats will then leave nasty sharp edges in the chambers. The rules state no chamber modification, so these edges cannot be removed afterwards. Besides interfering with the airflow, these edges will also create hot spots in the chambers when the engine is running, which will cause damaging engine pinking – or worse – destructive detonation.

Needless to say, these physically perfect castings are thin on the ground and finding a suitable one takes an awful lot of searching, not to mention the experience necessary to know what to look for in the first place – even if you're fortunate enough to have a stock of old heads to look through. Once a suitable casting has been found, it is then cleaned up and tested on the flowbench to see if the head actually performs to par (or preferably above average) in terms of airflow. All may look good with the ports but, as I've said before, looks can be deceptive and flow testing is the only way to be sure.

Once the final choice of casting has been made new guides are fitted: bronze alloy for the exhausts, standard cast iron for the inlets. Valve seat inserts are fitted for the exhausts (the bronze guides and inserts are used as part of the permissible lead-free conversion package) and high-flow three angle valve seats are cut for both the inlets and exhausts. Seat widths are usually 1.3mm with a minimum top and bottom cut. Once the seats are cut, any sharp edges remaining in the throat after the machining must be carefully removed by hand with a fine file.

Clean out the guides with some spray choke cleaner and push a small wad of tissue paper or rag through the guide to remove any swarf that may cause damage to the valve or guide itself. Then all that is required is for the valves to be lapped-in with some fine lapping compound. Only a few spins of the 'sucky stick' should be needed to give an even grey seat around the outer edge of the valve and on the head. Keep the pairs of valves for each chamber together for now, as the next requirement is to find the chamber volumes.

That is as far as the rules say you're allowed to go with these heads, except for skimming the head face to get the specified 36cc combustion chamber volume that's also given in the class rules, though some rules do not stipulate cubic capacities in the head. As to the amount of metal that should be removed by skimming, it is impossible to specify here, as the irregular shape of the combustion chambers means that no two heads are alike. This is where measuring the volume of each combustion chamber becomes necessary.

To measure the chamber volumes you will need a burette, a piece of perspex with a couple of small holes in (side by side) that's large enough to cover the chamber, and some grease. You will also need a sparkplug of the specific type which will be used in the engine. This is essential, as the sparkplug shape and design can make a difference to the readings. Install the sparkplug firmly into its hole, with a very light dab of grease on the threads before installation just to ensure a good seal. Place the head on some form of stand – a couple of wooden blocks will do – with the chambers uppermost. Lightly grease the inlet and exhaust valves around the seat and install them in the chamber to be measured. Then put a thin smear of grease on the head face around the outside of the chamber and stick the perspex onto the head. Don't allow too much grease to escape into the chamber from around the valves or from under the perspex. Fill the burette with water and zero the reading on the tube. You can use alternatives to water – paraffin, for example – but this demands far more care when using it; it's not very nice to your hands and, besides, water's more readily available, and spills don't really matter. You can now measure the volume of the chamber by carefully running the water out of the burette and into the chamber through one of the two holes, whilst allowing the air to escape from the other. Once the chamber is full, with all the air removed, the volume can be determined by the amount of water that has been used. Make a note of this reading. Empty the water from the chamber and give it a quick squirt of WD40 or similar water repellent to stop it going rusty. You will be surprised at just how fast corrosion strikes!

Do the same for the other chambers, remembering to use the actual valves that were lapped for those chambers, making a note of each chamber's volume as you go.

You can then use these measurements as a guide to how much material has to be skimmed from the head. A general rule of thumb for skimming the B heads is that 10 thou (0.01 inch or 0.025mm) of metal removal equals an approximate 1cc reduction of chamber volume. I must repeat that this is not a hard and fast rule, due to the previously mentioned variances between heads, but it can act as a guide to how much material to remove to achieve the desired chamber volume.

You will undoubtedly find that one of the combustion chambers is slightly larger than the others, usually either the chamber for cylinder 2 or 3. It seems to be a common occurrence on B heads, and for the blueprint specification heads it is this chamber that gets near spot-on for the 36cc by skimming. This may mean several further checks of the volume of this larger chamber are necessary, using the burette, in-between skims. These repeated checks are essential in order to avoid removing too much material and ending up with too small a chamber. Of course, with the larger chamber now reduced to the required size, this results in the other three chambers being slightly too small, with a volume slightly under the regulation size. As modification of the chambers is not allowed, the thickness of the inlet valves in each chamber is reduced by re-facing the seat, so making the valve margins thinner. The inlets have very thick margins as standard and thinning them down in this way does not

compromise strength or longevity, and, more importantly, the valve face remains unaltered. Yet again, this entails a lot of repeated reassembly and measuring of each of the chamber volumes with the burette, until all the chambers are equal to within 0.5cc. The valves are then number stamped so they can be replaced in the correct chamber if the head is ever disassembled at a later date.

All of this may seem a bit extreme, but it is vital if an engine is going to run smoothly and produce race-winning performance. This class of racing is intended to equalise the level of modification allowed to the engines, and so give good close racing. If you are competing to win, any legal means of improving performance has to be employed and this attention to detail is crucial.

The final process is to run a de-burring tool or a fine round file around the edge of each of the chambers to remove the nasty sharp frazes left by the fly cutter used for the skimming; the idea is to do the minimum of metal removal conducive with breaking those sharp edges. Again, this is to remove the chance of any hotspots forming, as mentioned before. A quick de-burr of the outer edges of the head face would not go amiss either, to save your hands from the nasty sharp saw tooth finish when carrying the head around.

The head will then need a thorough clean to remove every possible trace of grit and swarf generated during the work. Here, a hot wash tank is ideal, as this uses a mild caustic solution for cleaning, but water and a garden hose will suffice. Particular attention needs to be paid to swilling out the waterways and cleaning the small oil hole that feeds the rocker shaft. A vigorous poke about in the threaded holes for the studs and bolts with a small screwdriver will dislodge any

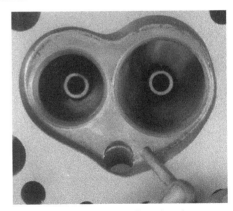

Deburring combustion chamber.

lurking grit – though chasing the threads with a proper cleaning tap is better. When thoroughly clean, a good spraying of WD40 or similar will chase out excess water and keep corrosion at bay.

As a brief aside, a mention of the method of skimming the head would probably be beneficial here. The best technique by far for metal removal from the head face is achieved using a fly cutter. This removes the material in a series of circular sweeps as it traverses the length of the head, leaving distinctive curved machining lines visible on the head face once the skimming is complete. This creates an excellent flat surface, but one that has a rough finish. The coarse surface finish is advantageous when using conventional composite gaskets as it enables the head gasket material to conform and compress into the tiny grooves, vastly improving the gasket's sealing ability and performance under combustion and other pressures. Skimming the head by any other means does not allow the gasket to 'key' into the head in this fashion. Obviously, with composite gaskets, the smoother the head's surface finish the poorer the seal will be, with resultant water leaks and eventual loss of power due to a blown gasket.

When using a metal (copper) gasket some people prefer a smooth,

Fly-cutting head face.

surface ground, shiny finish, as they fear combustion gases and fluids may track along the grooves made during flycutting causing failure.

If the metal gaskets do not come with any embossed sealing materials on them as standard, use of a thin smear of sealant (such as Wellseal or Permatex form-a-gasket) on both sides around the oil and waterway cutouts is highly recommended.

MODIFICATION WORK

The next type of head could be termed a road modified DIY head, as the results can be easily obtained with the minimum DIY work, coupled to good three angled seats and some valve shape modifications. From now on we are describing heads with the larger 1.625 inch (0.41mm) inlets fitted, using bronze guides and with exhaust seat inserts for the conversion to lead-free fuel specification. These modifications work equally well when applied to the smaller inlet valve heads – just ignore the seat diameters mentioned. The increase in inlet port flow due to these

modifications is shown as column E in Table 3 on page 42 (column D represents a standard port).

Now to the actual modifications themselves. If you look at the shape of a standard inlet valve you will see a raised lip above the valve seat; it may not seem like much but it does present quite an obstruction to the airflow. That lip needs to be removed to allow the air to flow more freely over the valve. To do this the inlet valve is cut back at 30 degrees to reduce the valve seat width to 70 thou (0.07 inch or 1.8mm). An additional 20 degree cutback then reduces the 30 degree cut to 60 thou (0.05 inch or 1.5mm). Both are blended into the back of the valve using 180 grit emery cloth. I use a professional valve re-facing machine to do this but the same result can be achieved using two drills (or one drill and an angle grinder if you're very careful not to remove too much material). Hold one drill in a vice and clamp the valve in its chuck. Put a round stone in the other drill and, with both drills running, use the stone to carefully remove the lip and re-shape the back of the valve slightly. Complete the job with some emery to give a smooth shape and finish to the back of the valve.

The exhaust valves only require the lip removing from the back, which is done with a 30 degree back cut that reduces the seat to 90 thou (0.09 inch/2.3 mm), or just a tickle with a stone as per the inlet valves.

The next part will require a visit to your friendly neighbourhood engine reconditioners. If they have the facilities to install the exhaust seat inserts and cut the three angle seats for the inlets and exhausts, leave written instructions as to how the seats are to be cut. For the head, the inlet valve seat should be cut to give a 10 thou wide (0.25mm) 30 degree top cut, a 50 thou (1.3mm) 45 degree seat – with an outside diameter

of 1.615 inches (41mm) – and a 40 thou wide (1mm) 60 degree bottom cut. This bottom cut can then be reduced later to around 5 thou wide (0.13 mm) opposite the sparkplug by careful blending using a stone in the drill (see diagram far right).

The exhaust valve seat in the head should have a 20 thou wide (0.5mm) 30 degree top cut, a 70 thou wide (1.8mm) 45 degree seat – outside diameter 1.33 inches (33.8mm) – and a 60 degree bottom cut. The throat is left as standard, though you can break any sharp edges left in the port where the insert has been fitted.

The combustion chambers will need to be worked on to remove any sharp edges or ridges left over from the inserts being installed and the new seats being cut. This is where some old valves will be needed to act as masks to protect the new seats. In order to do this they must be thinned down until they sit nearly flush with the chamber floor – the valve margin is actually reduced to almost a razor edge. These dummy valves can then be fitted to the head and will cover the seats while you take one of the large parallel stones and grind the port floor smooth. There is no need to try and get the port floor dead flat, that is something beyond the abilities of most DIY head porters (and some 'professionals,' too, it would seem!). All that's needed is the minimum of grinding necessary to remove the edges and ridges; any more and you will run into problems with balancing the chamber volumes when it comes to skimming the head.

For this specification of head getting the chambers to exactly the same volumes is not necessary, especially not to the extent of the standard race head described previously. Of course, you will still need to measure the volumes in order to work out how much to skim off, but as long as the chambers are within a couple of ccs of each other, that's fine.

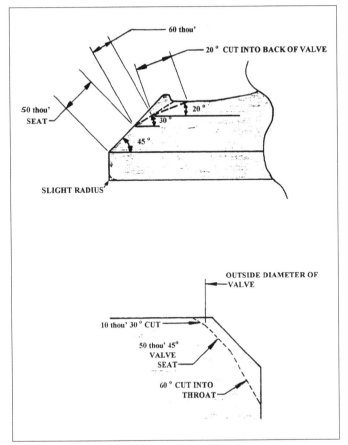

Inlet 3-angle valve and seat.

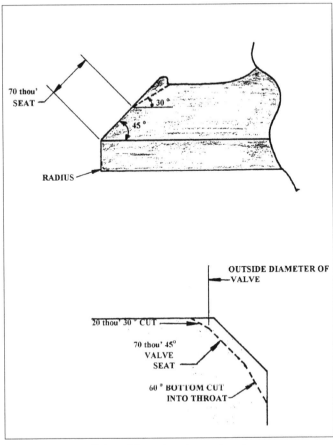

Exhaust 3-angle valve and seat.

Once skimmed, de-burr the edges of the chambers and head as described before and clean thoroughly ready for painting and assembly.

The topics of chamber volumes and compression ratios are covered later on.

The next stage of DIY head modifying involves more reworking of the ports and chambers. This type of head gives the flow figures shown in column F of Table 3 on page 42. It requires the bulleted guides as described earlier. For the inlet port the area around the valve guide is ground away to increase the flow around it (see above diagram). The port divide is sharpened at the split to aid swirl, and then blended into the throat and guide area. The remainder of the port needs to be tidied to remove

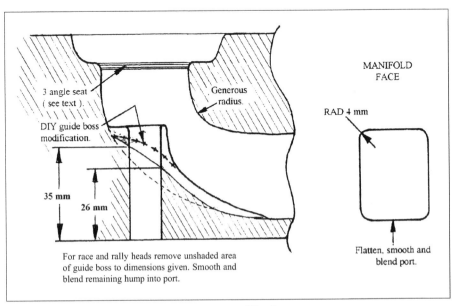

For race and rally heads remove unshaded area of guide boss to dimensions given. Smooth and blend remaining hump into port.

Exhaust port modifications.

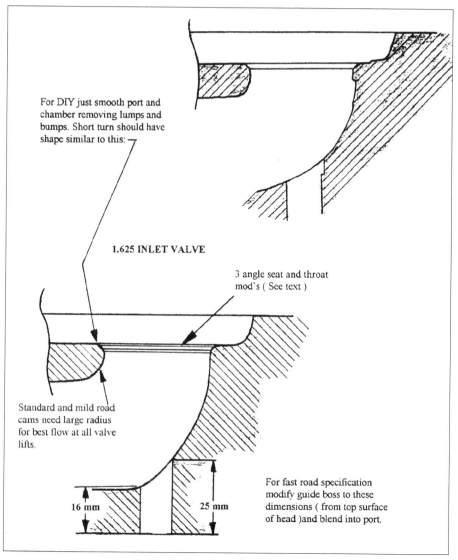

For DIY just smooth port and chamber removing lumps and bumps. Short turn should have shape similar to this:

1.625 INLET VALVE

3 angle seat and throat mod's (See text)

Standard and mild road cams need large radius for best flow at all valve lifts.

16 mm   25 mm

For fast road specification modify guide boss to these dimensions (from top surface of head)and blend into port.

Standard and modified inlet port.

Standard exhaust guide boss viewed from chamber.

Modified exhaust guide boss viewed from chamber.

lumps and bumps, but not enlarged, in order to maintain good gas velocity. The region near the manifold face (1 to 1.5 inches (25-38mm), where the port widens, benefits from careful smoothing and blending to remove the hump caused by the change of port shape – again, don't enlarge much, just smooth. The port should be around 31-32mm (1.2-1.25in) diameter at the manifold face. Remember to keep the stone moving at all times to avoid making

hollows, and finish off with a roll of oxide paper on the split rod.

The guide boss in the exhaust port is ground down to reduce the flow restriction it causes. The port is cleaned and blended into the corners, which are ground to a radius of 0.15 inch (4mm).

Once the guides have been installed, the seats can be cut to the specification given for the previous head, and the sharp edges carefully blended in to the modified throats.

Pressing in valve guides.

AIRFLOW (CFM at 25in H2O)				
Valve lift (thou)	D Standard 1.625 inlet	E Valve, seat & throat mods	F Modified port & chamber	G As for F with custom 1.625 inlet
50	17.4	20.7	21.6	21.5
100	37.4	39.8	42.6	42.5
150	55.5	62.1	63.9	64.7
200	73.3	80	83.2	84.4
250	88.2	94.2	98.3	100.3
300	98.6	105.2	109.4	113.4
350	108.2	116.4	120.3	124.5
400	113.3	121.9	126.8	132.3
450	117.1	124.9	131.5	137.7
500	119.2	128	135.6	140.4

Table 3: Airflow comparisons for various head modifications.

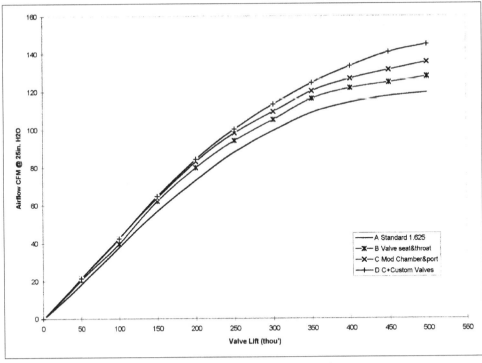

Airflow graph of various head modifications.

The chambers should be flattened and smoothed as before to remove sharp edges. If the head is intended for very fast road or sprint/hillclimb use, the region of the chamber wall adjacent to the sparkplug can be cut back at 7 degrees from the vertical to 1.2 times the valve diameter, or 0.325 inch (8.3 mm) from the edge of the inlet valve. Use a new head gasket as a template and scribe the bores onto the head face (a little engineer's blue may help highlight the marks). Don't open the chamber anywhere beyond this line or gasket problems will occur.

Column G of Table 3 above is the same port with a recut seat together with further throat reshaping and blending to suit a custom made 1.625 inch (41.3 mm) valve, with a better shape and a waisted stem, in place of the reshaped 1.625 inch (41.3 mm) original.

Over the years of tuning and modifying cylinder heads I have found that, with the introduction of unleaded petrol, the power producing quality of all fuels has decreased. From ongoing cylinder head refining, development and testing, it became apparent there was no need to open up the combustion chambers in order to unshroud the valves as on previous modified heads running standard valves. With a slight increase in compression ratio (to 9.75:1) we have found that the original

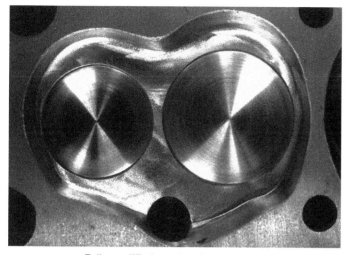

Fully modified combustion chamber.

Experimental combustion chamber for flowbench evaluation.

chamber shape produces similar power nowadays due to the extra squish area generating increased mixture turbulence and so burning the fuel better in the smaller volume chamber. The improved turbulence has outweighed the gains from increased airflow that unshrouding the chambers gave. Larger 43mm inlet valves also work well for road use without unshrouding if fairly mild cams are used. Unshrouding the chambers has become necessary only when large valves are used and more radical camshafts, as in rally or race applications.

RACE AND RALLY HEADS

To achieve optimum airflow through the larger inlet valves (1.69 in/43mm diameter and above), the chambers have to be unshrouded to a distance of 0.410 inch (10.4mm) from the edge of the valve, and angled to create a 7 degree slope (from the vertical) to the chamber wall. The point where the slope meets the chamber roof must be made a fairly generous radius or you may find fresh air! A waterway lurks beneath.

The size of the port at the manifold face is roughly 31-32mm in diameter for

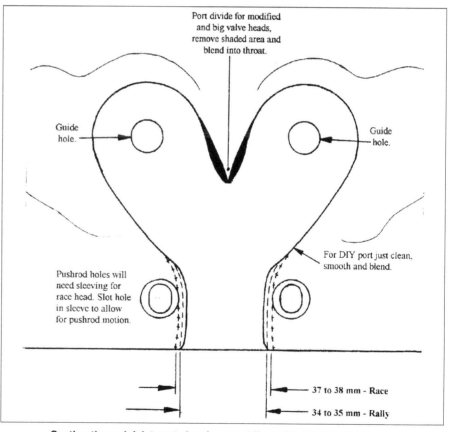

Section through inlet port showing port bifurcation and modifications.

Port divide for modified and big valve heads, remove shaded area and blend into throat.

Guide hole.

Guide hole.

For DIY port just clean, smooth and blend.

Pushrod holes will need sleeving for race head. Slot hole in sleeve to allow for pushrod motion.

37 to 38 mm - Race

34 to 35 mm - Rally

the modified head, once it's been lightly cleaned and smoothed. For the rally head this is opened up to give around

34 to 35mm (1.3-1.4 inch) diameter. For race use this is increased to 37 to 38mm (1.4-1.5 inch) diameter, the competition

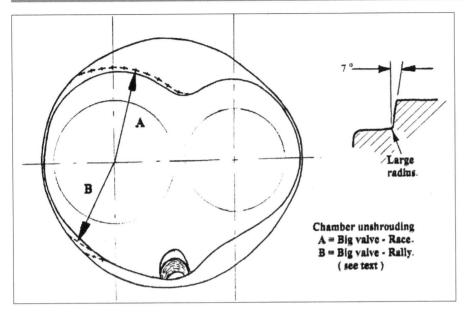

Chamber unshrouding
A = Big valve - Race.
B = Big valve - Rally.
(see text)

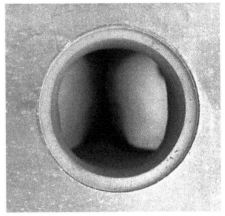

**Moss competition gasket over a
standard inlet port.**

**Dimensions for combustion chamber
unshrouding modifications.**

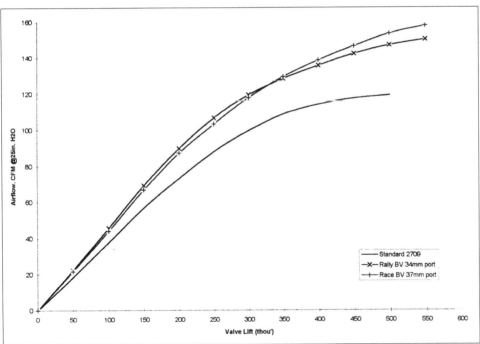

Airflow graph for competition heads.

inlet manifold gasket from Moss makes a good template to show how far to open it up. The remainder of the port needs working to blend in the enlarged section so there are no humps or bumps – aim for a near parallel tube from manifold

face to valve guide hole. Typical airflow figures are shown in Table 4, page 45.

With opening the ports out to this extent there is a danger of breaking into the pushrod tubes situated on either side. To overcome this problem the tubes

need to be sleeved and then drilled (slot them slightly in the direction of the pushrod motion) to take the pushrods. Use brass for the pushrod tube sleeves as it shows up plainly when you do grind through the port walls and you know when to stop working!

The port sizes need to be increased to allow for the greater volume of air required by the engine at high rpm (above 6000). This work doesn't show up as a significant increase in flow for the individual cylinders, but what has to be remembered is that both cylinders are drawing on the common port at the same time during the four stroke cycle (increasingly so with longer duration cams). Hence the need to open up the port at the manifold face to allow more volume of airflow; this is not a critical region as far as individual port flow is concerned. To use heads with these large ports for a road engine means that power below 3000rpm will be almost non-existent.

AIRFLOW (CFM at 25in H2O)			
Valve lift (thou)	Sprint/hillclimb 1.69 inlet	Rally head big valve 34mm port	Race head big valve 37mm port
50	21.7	22.8	21.7
100	43.7	46.2	44.1
150	65.6	69.6	66.9
200	85.3	90.5	87.7
250	103.2	107.2	103.9
300	116	119.6	118
350	124.8	128.6	129.6
400	133.4	135.8	138.9
450	140.8	142.4	147.1
500	145	148	154.1
550	147.5	150.1	159.7

Table 4: Race and rally head airflow comparisons.

COMPRESSION RATIOS

For road use it is probably best to stick to a maximum ratio of 9.75:1. For rally and sprint use this can be raised to 10.8:1 and for race use 11.5-12.5:1 (the large amount of overlap from the cams used serves to reduce the cylinder pressures).

These compression ratios have been established from many years of testing. If the compression ratio is increased much beyond these recommendations, severe detonation can occur. To stop the detonation the ignition timing would have to be retarded, and the result of not having enough ignition advance can be a hefty loss of horsepower! Compression ratio calculation information is given in Appendix I.

Even if you don't wish to have a go at modifying your own heads, I hope the foregoing will have provided a better understanding of the amount of work and skill involved in modifying cylinder heads.

Big valve rally port (top) and big valve race port.

VALVE SPRINGS

The subject of valve spring selection is covered at the end of the cam chapter, but the spring installed height can be checked during assembly of the head and is an important step in blueprinting a head. This is the distance from the spring seat on the head to the underside of the cap which has an effect on the correct operation of the spring as well as coil bind.

The recommended installed height is specified by the spring supplier. An incorrect height can be remedied by machining the spring seat on the head (increase the installed height) or adding shims under the spring (reduce the height).

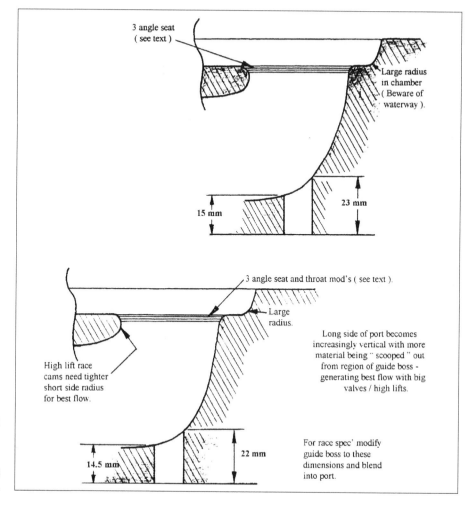

3 angle seat (see text)

Large radius in chamber (Beware of waterway).

15 mm   23 mm

3 angle seat and throat mod's (see text).

Large radius.

High lift race cams need tighter short side radius for best flow.

Long side of port becomes increasingly vertical with more material being " scooped " out from region of guide boss - generating best flow with big valves / high lifts.

14.5 mm   22 mm

For race spec' modify guide boss to these dimensions and blend into port.

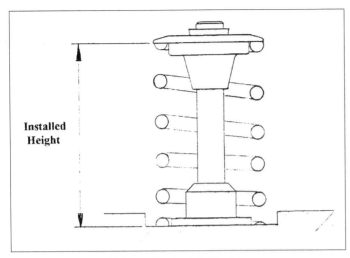

Valve spring installed height.

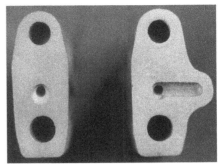

Inline and offset rocker posts.

To check for coil bind you need to measure the valve spring when fully compressed (be very careful!) and subtract this from the installed height measurement (including any shims used). The difference should be a minimum 50 thou greater than the maximum valve lift of the cam. If the spring binds before full lift is reached, valvetrain longevity will be severely compromised.

VALVE OIL SEALS

The standard rubber O-ring seals are ideal for all road-going applications. When a head has been converted to run on lead-free fuel and has bronze exhaust valve guides, I usually omit the seals, fitting them to the inlets only. The bronze guides run tighter clearances than the cast iron type, so oil control is better. Any extra oil down the guide helps lubrication before it's burnt and expelled with the hot exhaust gas, so no contamination of the inlet charge occurs.

THE ROCKERS ARMS & ROCKER SHAFT

Any wear on the rocker shaft or slop in the rocker arms is going to upset the timing of the valve events and lose power, and using an old shaft on a newly assembled engine is silly. The shaft should be dismantled and checked for any wear ridges at the rocker positions (usually on the underside of the shaft), and where the posts clamp the shaft (the shaft actually frets inside the posts). Use a micrometer for absolute accuracy, or run your fingers along its length. Any wear at all means the shaft should be replaced (complete, reconditioned rocker assemblies are not expensive); just ensure the correct oil feed is specified, in-line or offset, to suit the head (see photo above right).

You can recondition the shafts yourself by using an oversized shaft, but the rocker arms and the posts will need to be accurately reamed to suit the larger diameter.

Any wear on the rocker arm pads that contact the valves should also be removed, especially if you are changing the cam to a performance profile but retaining the original rockers. These wear ridges can cause side loading on the valve and accelerate wear.

Anyone with the earlier aluminium rocker posts should change to the stronger and more rigid steel type before any engine uprating is considered.

To blueprint the valvetrain the head must be fully assembled and fitted to the engine, and the tappet clearances set. By slowly turning the engine in the direction of rotation, check that the middle of each rocker arm pad is centralised on its valve tip at half maximum valve lift, achieved by putting thin shims between the pillar and rocker or removing material from the side of the rocker.

If the head has been skimmed the pushrods may have to be shortened (machine the round pads at the bottom of the pushrods, keeping the same original shape) to achieve this. If the head has not been too heavily skimmed you may be able to achieve the same result by putting several thin shims

Something of an extreme example, this rocker shaft exhibits both wear from the rocker arms (visible above the oil holes) and rocker post fretting (the rippled surface in-between).

(as found under the end pillars of later heads) under the rocker pillars.

All valve lifts must now be checked and equalised. Try to match all the others to the one with the highest lift (or to the correct specified lift if racing class rules apply).The pads on the ends of each of the rockers may need to be slightly reshaped, so all the valve lifts are nearly equal (within 5 thou is close enough).

You can see it takes an incredible amount of time to check and adjust everything until it is within optimum specification: all this would be a pointless exercise with a worn rocker assembly.

HIGH RATIO ROCKERS

Special high ratio rockers are available for the B series engine. These give a 1.625:1 ratio rather than the original 1.42:1, an increase of some 14.5 per cent. These rockers are made from aluminium alloy and have needle roller bearings for the pivot on the shaft and a roller tip to operate the valve. The only problem I can see with this is it's doubtful whether the needle rollers are an improvement as the

rocker movement is reciprocating rather than rotary. Therefore, the loads are taken by a couple of needle rollers on the underside of the rocker and are greater due to the small contact area of each bearing. These rockers become sloppy with use as the rollers and shaft wear quite quickly.

Rockers using plain bronze rocker shaft bushings are available. These are made by Harland Sharp in the USA and supplied as rocker arms only, to keep cost down. You can re-use the original rocker posts, shaft and adjusters if in good condition.

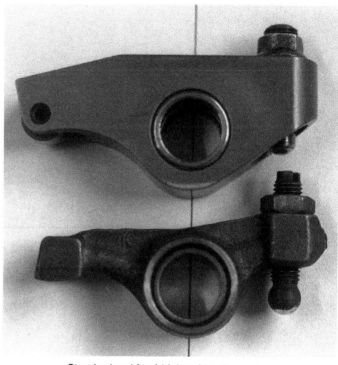

Standard and (top) high ratio roller rocker.

High ratio rockers don't alter the length of time the valve is open or closed, but do increase the net lift and acceleration of a cam, ie, at any time

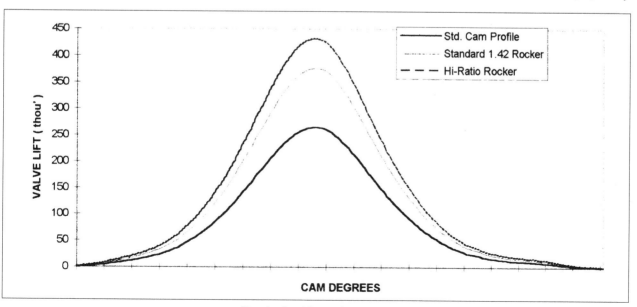

Standard vs high ratio rockers.

the valve is lifted just a little bit more with high ratio rockers. The valve springs must be checked for coil bind at full lift (a minimum 10 thou clearance between each coil of a spring is advisable) and changed if necessary. You also need to ensure that the exhaust valve will not hit the block at full lift, so a bore cutout is necessary if there isn't one already. The rocker (tappet) clearances will also need increasing (multiply by 14 per cent).

Heads and blocks that have been skimmed, valve seats re-cut, different thickness head gaskets, and re-ground cams with smaller base circles, means rocker geometry can be upset. Adding roller rockers is not as straightforward as first thought, and the installation requires a little effort and time to ensure all is well when fitting.

High ratio roller rockers – it's a pity to have to hide them under the rocker cover.

It is essential to check for correct geometry. Ideally, the tip of the rocker arm should always push on the exact centre of the valve stem, which reduces side loading on the valve stem and minimises valve guide wear. In reality, this seldom occurs, as the rocker tip describes an arc in motion. All that can be done is to try and keep the rocker arm as close as possible to the centre of the valve tip throughout the entire camshaft lift cycle. With the valve closed, the roller tip needs to be just off centre of the valve tip on the side towards the rocker shaft. During the opening cycle it should roll over the mid point and travel slightly towards the outside edge (away from the rocker shaft) of the valve tip at full lift, then return to its starting position – the roller being centred on the valve tip when at mid-point of lift.

Mocking up the valve train with black marker pen applied to the valve tip is a way to show the sweep of the roller after running through a full lift cycle. The valve tip will be clean where the roller has been. Adjustment can be made to the rocker shaft shims (up to 3mm thick shims is a start point) or pushrod length, if using an adjustable for checking, and the contact pattern re-checked until optimised.

For all the valves, check that the pushrod cups do not foul the rocker body where they meet the adjuster ball ends, and verify the pushrod clearances (easily rotate them with your fingers) to ensure they do not stick or rub in the cylinder head throughout the entire valve lift cycle. It will mean manually turning the engine over a few times whilst inspecting all positions. The rocker posts may need to be raised with shims and the pushrod holes possibly require some clearance grinding to accommodate the change. Alternatively, an adjustable checking

High ratio roller rockers with plain (not roller) shaft bushings.

pushrod will have to be acquired or fabricated, then used to set the correct rocker geometry. By lengthening or shortening the adjuster the rocker sweep can be checked. When correct, the measurement from the adjustable pushrod is used to have custom ones made to the correct length.

With a standard cam, fitting high ratio rockers gives an immediate 10 per cent power gain throughout the rev range.

A couple of examples of engines using high ratio rockers are shown in the section Power Recipes 2.

GASKETS

Choice and fitting of head gaskets is covered towards the end of chapter 6.

THE HRG DERRINGTON CROSSFLOW HEAD

This alternative head for the B series engine is in aluminium alloy and has four separate inlet ports on the sparkplug side of the head. The ports are narrower than those of the cast iron heads, giving

Inlet ports on Derrington crossflow head. Exhaust ports are situated as per cast iron head.

Derrington crossflow head combustion chambers.

A complete HRG Derrington setup.

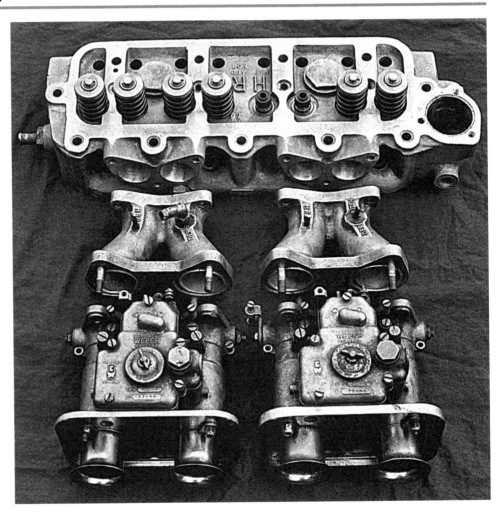

higher gas velocities at mid range rpm. This increases the engine mid range torque but, unfortunately, limits the alloy head's ability to make horsepower, to less than a standard cast iron head. In modified form the maximum bhp is the same as for a well modified cast iron head, but the Derrington generates around 15 per cent more mid range torque.

Typically, compression ratios need to be higher than those for the cast iron heads because the aluminium is more efficient at heat removal. The high thermal efficiency removes heat from the combustion chamber, heat that could be used to make power, so the remedy is to increase the compression ratio. This raises the combustion chamber temperature (and hence bmep) to the levels of the iron heads.

Typical compression ratios are 10.5:1 for road use, 10.8:1 for fast road and 12-13.0:1 for rally and race use.

AIRFLOW (CFM at 25in H2O)			
Valve lift (thou)	Standard head 1.56 inlet	Fully modified head 1.56 inlet	Fully modified head 1.625 inlet
50	17.1	19.4	21.4
100	35.5	41	42.6
150	53	61.3	63.5
200	67.5	79.7	82
250	78.5	93.5	96.2
300	83.8	104.5	109.1
350	86.9	112.8	117.4
400	88	115.8	123.6
450	88.8	119.7	127.7
500	89.7	123.1	132.4

Table 5: HRG Derrington alloy head airflow comparisons.

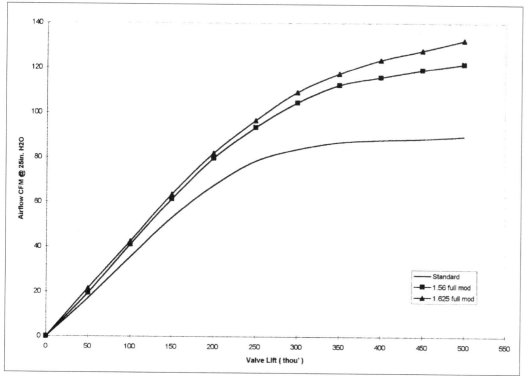

HRG Derrington crossflow head airflow graph.

MSX AND ALLOY 5 PORT HEAD

There are a couple of alloy cylinder heads for the B series available from the aftermarket: a 5 port replica with a PMI identification mark, having 1.56 inch inlet and 1.34 inch exhaust valves, and available drilled or undrilled for air-injection; and also a 7 port conversion (4 inlet ports, 3 exhaust) using the same valve sizes, with an MSX identification on the casting. The flow figures can be seen in Table 6 on page 54.

The quality of the 5 port heads is poor in standard form, generally being not well finished.

Slipped/non-round cores for the inlets

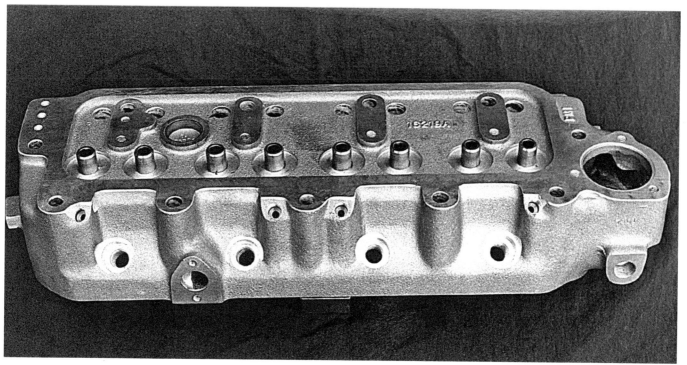

Alloy 5 port head.

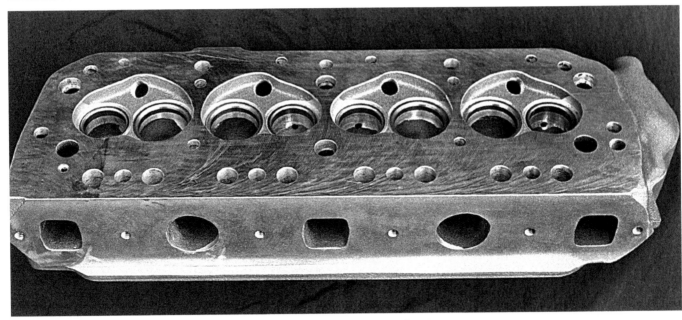

Alloy 5 port head chambers and ports.

necessitates creating larger than standard ports to attempt to true them up, which means the inlet gasket and inlet manifold have to be enlarged too. The short side turn of the inlet ports is very tight and doesn't allow as large a radius as with a cast iron head. That said, in modified form they are nearly as good as modified cast iron heads. Ultimately, the inlet valve size is restricted to a 1.69 inch maximum because of the need for both inlet and exhaust valve seat inserts; the diameter of the exhaust insert precludes a larger diameter inlet seat.

The 7 port conversion is available from Webcon in the UK. As standard the heads are somewhat of a disappointment, the machining work is incomplete, plus having poor seat dimensions and chamber walls too close to the valves for good flow, especially around the inlets. Poor casting of the head ports and those in the inlet manifolds necessitate a lot of rectification

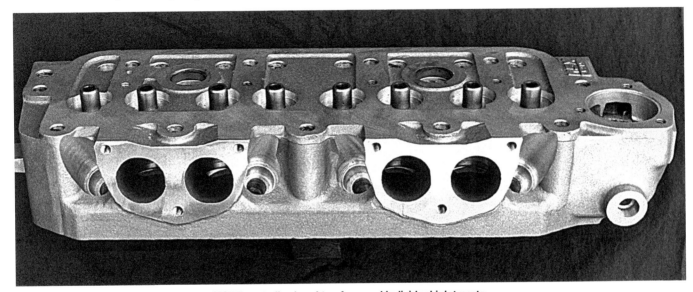

MSX 7 port alloy head top face and individual inlet ports.

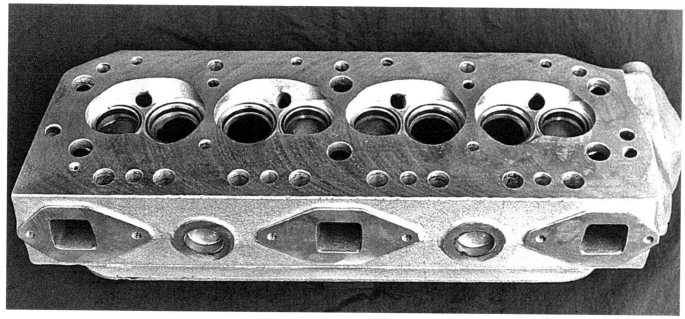

MSX 7 port head chambers and exhausts. Cylinders 2 and 3 are Siamese.

work. The head gasket face should also be checked to see if it's warped, and skimmed as necessary. After considerable alterations to ports, seats, valve throats and chamber walls the modified head makes very strong mid range and good top end power. The twin Weber setup should be run for best effect. With the inlet manifold and carbs installed there will be access issues for the distributor, with possible difficulty getting at the oil filter, too.

When modified the MSX heads return excellent power, however purchasing them and having the necessary modification work does not come cheap, and while used cast iron heads are available, these heads are more for a niche market.

Testing an MSX head on the flowbench.

AIRFLOW (CFM at 25in H2O)			
Valve lift (thou)	Standard MSX 7 port 1.56 inlet	Modified MSX 7 port 1.625 inlet	Standard 5 port 1.625 inlet
0	0	0	0
50	17.9	20.33	17.56
100	37.16	42.81	38.74
150	56.51	60.24	57.6
200	72.85	76.25	73.28
250	86.0	91.46	82.24
300	95.25	105.56	88.24
350	99.57	118.51	92.75
400	101.5	127.76	96.0
450	103.2	130.59	97.95
500	-	129.83	-

Table 6: MSX head flow figures.

MSX 7 port with twin 45 DCOE Webers.

Chapter 6
The bottom end

As always, when building anything the end result is only as good as its foundations, and the same is true, if not more so, when engine building. The cylinder block is the engine's foundation, and preparing it correctly demands great care and attention to detail. This necessitates a lot of time and considerable amount of money, as the necessary machining and rectification work will have to be done by a skilled machinist with the correct equipment. There is no cheap and easy route to perfection!

The goal with block preparation is to get all the component clearances within specification, so they all work together in harmony with each other and their surroundings, and thus reduce frictional losses. To do this it's necessary to measure and check everything. You will need to obtain a very comprehensive workshop manual – the factory original is best – and access to measuring instruments. A cheap engine stand will

also pay dividends, making the engine work much easier.

It is a good idea to wear latex or nitrile gloves during the initial disassembly and cleaning, or use barrier cream on your hands, as waste oil is not good for your skin.

The following text details the full engine blueprinting process, and you can apply some or all of it to your engine build, as you wish.

STRIPPING AND CLEANING

When starting from scratch with a complete engine, check for any external signs of water or oil leaks. If possible, drain the sump contents into a container through an old sieve to catch any bits that may warrant further investigation later. If this is not possible, check for any debris in the sump once it's removed; a useful indicator of the state of the engine.

Dispose of the waste oil in a sensible manner at an approved

recycling site, as polluting is a serious offence.

Most manufacturers number their main and big end bearing caps but, if yours aren't, be sure to do this yourself. With the engine disassembled, everything that is to be reused – crank, rods, etc – needs to be cleaned well enough to allow measurements to be taken. The crank and rods can be given a spell in a hot caustic tank to remove the old oil and varnish accumulated during service.

CYLINDER BLOCKS

The question of which is the best block to go for is largely irrelevant except for racing, as most people wish to recondition the original engine or have a replacement of a similar kind. If you're interested, though, the best block to go for is the 1971 to '74 18V chrome bumper type with five main bearings and bolts for the crank main bearing caps. Some earlier blocks used studs and have weaker caps. The pre-18V and early 18V blocks will not fit in rubber bumper cars, however, as the engine mounts don't fit on the exhaust side of the block – on genuine rubber bumper engines the oil gallery is shorter on this side of the block to allow for a repositioned mounting boss.

To help remove years of corrosion from the waterways you can blank off the water pump holes (leave the water pump fitted and block the holes), and fill them with a concentrated mix of central heating radiator de-scaling fluid, leaving it for a reasonable length of time to do its stuff before draining.

The block should have everything removed: core (freeze) plugs, drain plug and cam bearings. The three brass plugs at the back and two at the front of the block can be drilled out to allow access to the oilways if you wish. These plugs are available from main dealers – or make your own replacements. An alternative would be to drill and tap the block to allow threaded pipe plugs to be used for blanking purposes.

The block will then need a good long soak in a hot tank of caustic solution to loosen the muck from the oil and waterways. A common problem here is that the passage behind the screw-in drain plug is usually blocked solid. A good dig about in the hole with a long taper punch or screwdriver (a hammer assist may be called for, but use it gently) and down the connecting passage from the top face of the block, will be required to clear it.

Once out of the caustic solution the block will need a wash and coating with a thin oil or water repellent to keep corrosion at bay. When thoroughly clean give the block a visual check for any damage or obvious cracks. The blocks are mostly trouble-free apart from the expected bore wear, which

Brass plugs (arrowed) need to be removed in order to clean oil galleries.

The end of an old pushrod has been ground to form a chisel ...

can be felt as a ridge at the top of the bore. If possible measure the bore size to determine the next oversize to go for, or you could decide to go larger to gain more torque from a capacity increase.

Block preparation and machining

Now comes the vital part of block preparation: ensuring that the bores are square and perpendicular to the crankshaft. While they may have been accurate from the factory (doubtful!), over time the thermal cycling of the engine relaxes any stresses there may have been in the block casting and it can distort. Previous reconditioning work can exacerbate the problem. This distortion needs to be rectified to create a powerful, free revving engine.

Using a straight edge and a depth micrometer with a ball end adapter (or a ball bearing), measure from the block's sump face down to the middle of the main bearing arches and note the readings. Then, with a short straight edge against the top of each of the bearing arches, use a long extension measure from the head face down to the straight edge (a spare pair of hands is useful here!). Again, note the readings; with them you should be able to see if the crank is parallel to the bottom or top surface.

Next, fit the main bearing caps and torque to specification in order to measure them for correct diameter and any ovality or taper. You may be lucky and have dimensionally accurate main bearings, but any problems here will need correcting by line boring. To do that, a datum surface to work from is required.

If your earlier measurements show the crankshaft is parallel to the head face of the block, the sump face will need skimming so it, too, becomes parallel. The block can then have any rectification work necessary done to the main bearing bores – the block is usually positioned head face down for the work to be done, so the end result will be parallel to both block faces.

If measurements show the crank is parallel to the sump face then the head

... to create an excellent means of clearing the corrosion in the waterway accessed from the block face, by patiently tapping firmly whilst rotating ...

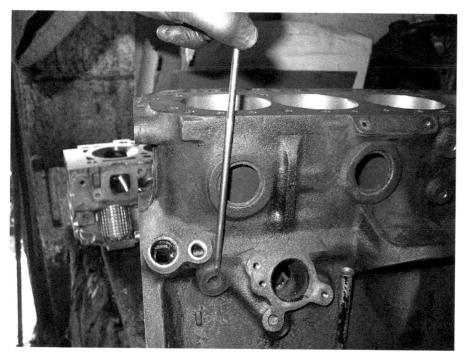

... until it can be seen through the hole for the drain plug.

Measuring main bearings for ovality.

Measuring bores.

face needs skimming to make it parallel prior to any align boring of the main bearings, for the above reason.

Align boring involves re-facing the cap mating faces, assembling the now undersize big end caps on the block and re-honing them to the correct size.

If both block faces are out of line with each other and the crank, put the block head face down to be skimmed. Measure from the main bearings to the table, using the straight edge, and put shims under one end of the block to bring them level. Securely clamp the block down and re-check. When correct skim the absolute minimum necessary from the sump face to bring it parallel. Turn the block over and minimum skim the head face.

Once all this squaring and truing-up has been done the block can be bored to size. The boring bars that sit on the top of the block are OK for normal reconditioning work, and, in the hands of a skilled machinist, can work well preparing blocks for performance applications, provided many light

cuts are made when opening out the cylinders. For high performance engine preparation Ideally what is needed is a large, stand alone, boring machine that has its machining axis perpendicular to the bed the block sits on. This will ensure the finished bore

is perpendicular and square to the crankshaft centreline.

Take the new pistons with you to the machine shop so that each bore can be sized accurately to each piston. Just boring and honing to a 'universal oversize' dimension without reference to each piston's specific dimension is not good enough for this level of application.

The final requirement for an accurate bore size and finish is fitting a

Measuring pistons.

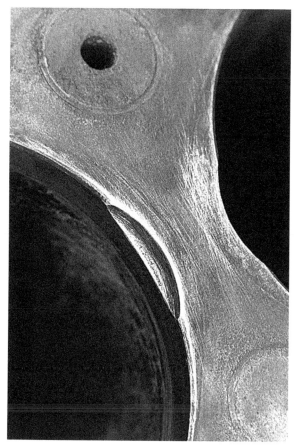

Exhaust valve cutout in cylinder block.

fine grit impregnated cork pads are used to knock the tops off the honing finish – provides a means of getting a quicker ring seal, but is not an essential procedure. Ask for a fairly large chamfer to be added at the top of the bores once the boring is complete. Besides offering a lead-in for the piston rings during installation, it also allows a marginal improvement in airflow by unshrouding the valves.

Don't forget the block cutout to clear the exhaust valve (about 120 thou/3mm) if there is not one already, or if you are going to use an 18V head on a non-18V block.

Give the block a good clean. At some stage before final cleaning you will have to do a dummy build to measure how far the pistons are down the bore in order to know how much to skim off the block to achieve the desired piston height – and hence compression ratio.

Some piston manufacturers make pistons for large bore conversions that have reduced pin to crown height (they are shorter) to compensate for the large increase in bore size. It can be a lot easier to leave the piston slightly further down the bore rather than go to the bother, and expense, of trying to find extra ccs in the cylinder head combustion chamber to get the desired compression ratio.

With capacities of up to 1868cc (+ 60/1.52mm), I like to aim for a piston fitted height of 12 thou down the bore (as per the standard race class rules). With larger capacity increases you will need to calculate the required volume above the piston at TDC (including chamber capacity) to achieve the chosen compression ratio. The calculation is at the back of the book.

Everything must be scrupulously cleaned again upon disassembly.

torque plate to the block prior to boring and honing. A torque plate is a thick piece of steel, intended to simulate the stresses in the block caused by the fitting of the cylinder head. As the head bolts extend into the block alongside the bores, distortion occurs as the block flexes when the head bolts are torqued. This pulls the bores out of round, so weakening the piston ring seal and causing loss of power. The torque plate helps ensure that the bores are round when the head is finally fitted.

The block should be bored to within a few thou of the finished size and the remainder honed out. A good 45 degree crosshatch finish to the bores is required, for good oil retention and piston ring sealing. A plateau hone – where special

Chamfering main bearing saddles.

Finishing and cleaning the block

With all the machining work out of the way the block needs final cleaning. A beneficial touch is to fully deburr the upper and lower surfaces and any edges inside the block, especially after machining. Deburring and radiusing can also aid oil drainback in some cases, and help eliminate potential stress raisers where cracks may begin.

Get a selection of stiff bristled cleaning brushes (available from trade suppliers) to thoroughly dig around and poke through all the oilways and passages until they are spotlessly clean.

All threaded holes that receive a fastener should be lightly chamfered to remove any high spots and stop the threads from pulling out when the studs or bolts are tightened to specification. They should also be chased with the correct sized tap (carefully, to avoid removing any metal) to clean up the threads. Use a bottoming tap on blind holes and clean all the way to the bottom. Once clean use the small brushes to remove the debris – or blow it out with an airline but wear safety goggles.

A cooling modification applicable to non-18V blocks is to enlarge the rearmost waterways (to 14.29 mm/0.563 in) – an old special tuning tweak which improves coolant flow to the cylinder head and around the hotter running No. 4 cylinder.

Chamfer all around the main bearing saddles with a fine file to remove any imperfections that may stop the bearing shells or big end caps from seating correctly. The cap parting line can be chamfered by drawing along a fine file.

Thoroughly scrub and clean the block to remove all traces of dirt or swarf. Hot water, stiff brushes and washing powder work well, but keep

Core plug straps provide insurance against them working loose over time.

the surfaces wet at all times otherwise corrosion will occur, instantly! A compressed air source is useful for drying the block, or spray with a water repellent (like WD40) or light oil to chase the moisture away. When clean, dry, and corrosion-proofed, install the new oil gallery plugs if they were removed. The bores should then be cleaned by wiping with clean paper rags and ATF (Automatic Transmission Fluid) until totally spotless (the fine grit and swarf from the honing process lurks in the crosshatch and is very stubborn, resisting most attempts at removal). You should be able to wipe a clean paper rag down the bore and it remains clean. Leave a thin coating of oil on the bores as protection.

New core (freeze) plugs can be fitted: a thin smear of sealant helps prevent any leaks. To prevent them working their way out of the block over time we now fit them all with retaining

straps. Holes for 4mm hex head cap screws are drilled and tapped into the thick casting either side of the openings to attach the flat aluminium bar that keeps them in place.

When all is done seal the finished block in a plastic bag for safekeeping until it's time to assemble the engine.

PISTONS AND OVERBORES

The largest oversize (in thou) I recommend is to + 60 pistons. The factory originally didn't recommend going over + 40 on a re-bore due to core misalignment and poor quality with the earlier castings that could lead to porosity problems (almost breaking through to the waterjacket). I have not encountered any problems going out to + 60 and this size also leaves room for a further re-bore, to + 80, in the future.

Going even larger with the bore size means the block is eventually scrap or

Engine capacity table		
Overbore (thou)	Bore & stroke	ccs (*big bore)
Std.	80.26/88.9	1799
+20	80.77/88.9	1822
+30	81.02/88.9	1833
+40	81.27/88.9	1845
+60	81.78/88.9	1868
+80	82.29/88.9	1891*
Lotus TC Std.	82.55/88.9	1903*
+40 TC	83.57/88.9	1950*

will need liners fitted to reduce the bore size should a future rebuild be needed. The other problem with very large bore sizes, apart from the possibility of finding a waterway (block scrap or needing liners) is that the remaining bore walls have become so thin they flex. This loses power through inefficient piston ring sealing and causes increased oil consumption (characteristic of some big bore conversions), with it finding its way past the struggling oil control rings and being burnt.

A three to three and a half thou piston to bore clearance (total) is recommended, the latter being for race engines as well.

As good engine blocks are now becoming scarce, a worn example can be saved using liners, provided it is in otherwise good condition. This means a block can be returned to the original factory bore size, if so desired. Liners are made from spun cast iron, and are good for piston ring seal and hence power, plus they help restore block rigidity lost following multiple or large overbores.

A 2.5 thou piston to bore clearance (total) is recommended for road use, and 3 thou minimum clearance for race.

The piston rings should be end-gapped as given in the manual and the ends of the rings lightly radiused with an engineer's stone to break the sharp edge.

CRANKSHAFT

With the standardisation of parts, the early B's steel crankshaft was replaced by a cast iron one. Surprisingly, the cast crank is the superior of the two, being better balanced and with better wear characteristics.

The best cast crank is the flat sided one to be found in 18V engines from 1971 to '74 (including the Morris Marina to 1974) which is fitted with a removable spigot bush because of

A professional tool such as this speeds up the ring gapping process and ensures accurate straight and square edges.

the smaller diameter input shaft in the Marina gearbox). The later Marina crank has large rounded counterweights and the wrong size spigot bush. In order to use these in a B it's necessary to have the end machined out to take the larger bearing for the input shaft.

The crankshaft should be cleaned in a hot tank of caustic solution, after which it should be crack tested before any further work is done. It's not that there are any problems with them, but more for peace of mind. The crank webs can be deburred with an angle grinder (make sure any race class rules permit this!) to remove casting flash and eliminate stress risers. Protect the bearing surfaces with thick masking tape or duct tape to prevent damage from flying swarf.

If you are fortunate and have an engine that has been well looked after, having had regular oil changes, the crank may not need any machining work done other than a polish of the journals with some fine grit paper.

Here's how to check the crank. To check for a bent crank mount it on vee blocks or the main bearings in the block

Crankshaft types with earlier, flat-sided type at bottom.

and check each main bearing journal with a dial type indicator for run out. Any run out can be remedied with a press. If you are not sure about doing this, ask the machine shop to do it for you.

Forged steel cranks can be straightened, but cast iron ones cannot and are best rejected.

Measure the crank journals at various places along and around the

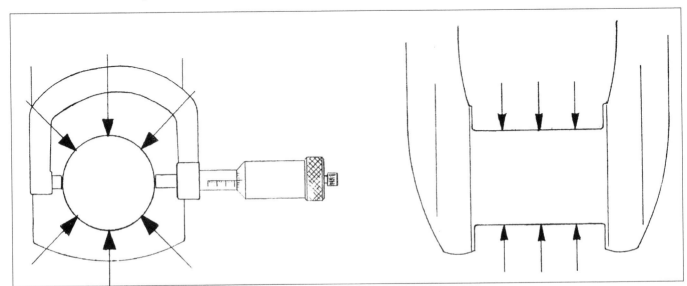

Measuring crank journals for taper and ovality.

Measuring the crankshaft journals.

Once grinding is complete the oil supply holes can be lightly chamfered – just enough to remove the sharp edges left after machining and aid oil flow out to the bearings.

Thoroughly clean out all the oilways and clean and chase the flywheel bolt holes with a bottoming tap and that's the crank sorted!

CONNECTING RODS

There have been several different connecting rods used over the years. The very early MGB 3 bearing engine (18GB) had the angled parting line for the big end caps and floating gudgeon pins (little ends). They were carried over from the 1622 MGA, where the caps were angled to make them narrower in order that the rods would fit down the MGA's small bore and over the crank journal. MGB 5 bearing engines up to 1969 also used angular split rods (identified by raised 12H1019E on shank).

Next (18GB and early 18GG to 1971) came rods with the more

bearing surface with a micrometer. The measurements around will detect any ovality and those along the journals will show up any taper. If either is present on the mains, big ends or both, or if there is wear outside the tolerance range specified, it will need to be reground.

If there is no visible damage to the crank thrust faces ask the machinist not to touch it as there is no need.

The crank should be reground to the bottom limit of the next undersize, and ask for a generous radius between the journal and the web which, again, eliminates stress and strengthens the crank. A crank ground with a square or sharp corner is more likely to fail at this point.

Before grinding ask if the reconditioner can correct the stroke as well. If not, look somewhere else as not every crank grinding machine can do this. The stroke correction means that all pistons should reach equal heights in the bore (as long as the con rods and pistons are the same), rather than, upon engine assembly,

one being slightly lower or higher than the rest. Equalisation of each cylinder's compression ratio is made easier, so the power produced per cylinder is similar, giving a smoother running engine.

MGB connecting rods (from left to right): early, 18V and late 18V.

conventional horizontal parting line big end and a small end bush for a floating gudgeon pin (12H444E). These rods have large balance pads at each end and are rather heavy, which can be remedied by machining. These are the rods to use if you are going to use Lotus twin cam pistons for a large bore conversion, as they have the correct little ends for the floating gudgeon pins. The other alternative is to have the little ends of the other rods bored out to suit.

The final type (18V) has a press fit gudgeon pin and the conventional horizontal parting line. The very late rubber bumper engines have rods that haven't got the big balance pads top and bottom; they were supplied balanced from the factory in selected sets of four. The pistons for these will have to be removed and fitted by specialists with the correct equipment to avoid damaging or distorting the piston, pin or connecting rod.

Checking the con rods

After making sure the block is straight and true, the connecting rods need looking at to ensure that they do not send the piston up and down the bore all cockeyed. The check for straightness uses a special alignment rig, so you will have to find someone to test them for you. The alternative is to use mandrels and vee blocks as detailed in the workshop manual. Any twist or bend is usually cured with a long bar and the application of muscle power. What also needs to be checked is the big end bearing housing for ovality and taper – with a bore comparator or internal micrometer. This is done with the con rod bolts torqued to correct specifications. Any problems will mean having the big ends re-sized to ensure correct bearing crush and fit. Misalignment of the bearing caps means the rod(s) will need replacing.

The little ends on those rods that have floating pins just need checking that the pin slides freely in its bush.

Contrary to popular belief, tuftriding the con rods serves little purpose. Polishing and shot peening can be done if desired for race use, but is not essential.

What is essential are good quality bolts. The standard ones are fine for all stages of road use, but uprated ones from the likes of ARP offer much greater strength and peace of mind for high rpm or race use.

Connecting rod end float will need checking when test fitted to the correct crank journal – the tolerances given in the manual are ideal.

Checking big end bearing housing.

Use a glassplate and fine wet and dry to thin the big end shoulders if necessary.

All the edges and parting lines should be lightly chamfered with a fine file, as per the main bearing caps and saddles, to ensure bearing shells and caps will seat properly.

BALANCING

Any changes of or to components – such as a crank regrind or oversize pistons – will alter the way they interact with each other and affect the forces acting upon them. A smooth-running engine that doesn't feel harsh or rattle must be balanced.

The pistons are balanced by equalising them all to the lightest one (within 0.5g is accurate enough) using scales. Remove material from inside the gudgeon pin as this is heavier and doesn't affect piston strength.

Experience has shown that the factory-balanced B engines are well balanced enough even for racing, and equalising overall weight rod to rod is sufficient when used in combination with balanced pistons.

The connecting rods can be balanced in a similar manner, but they do require a special fixture to allow only the little ends, then only the big ends, to be balanced. Again, they're equalised to the lightest rod by removing metal from the

Balancing pistons.

balance pads at each end. This should be done before the rods are shotpeened, if that is the intention, or the good work will be undone.

Some race classes stipulate that one rod must remain standard to eliminate any undue lightening work or modification. Obviously, in this case, the lightest rod must be selected and the rest equalised to it.

A specialist is needed to balance the crank as this is done dynamically with the crank spinning. Weight is removed from or added to the crank webs until any out of balance forces have been removed or reduced to acceptable levels. Once the crank is balanced the front pulley is added, then the flywheel and finally the clutch assembly. This is so that if the clutch is changed at some stage the whole lot will not become out of balance, though clutches are well balanced anyway.

Radically altering the basic shape of the factory crank by knife-edging or wedging the webs can lead to harmonic vibration in use and consequent bearing or catastrophic crank failure.

Don't be tempted to modify (lighten) the front pulley/vibration damper. Any changes to its mass will drastically alter damping characteristics (the frequencies it is meant to damp). This may lead to the damper adding to vibration problems rather than alleviating them. If you wish to lighten the pulley, change it altogether for one without the damper.

CAMSHAFT AND CAM DRIVE

The cam bearings will have to be installed by an engine reconditioner, unless you can fabricate some means of pulling them into place in the block (drifting them in with a hammer can distort them). Either way, check to ensure the oil holes in the bearings are lined up with the feed holes in the block.

Duplex chain and timing gear.

When installed you can carefully slide the cam into the block, without any oil, and rotate it by hand to check for any tight spots. If there is any binding remove the cam and check each of the bearings, the interference point will show up as a shiny spot. This bit can then be lightly polished with some very fine wet and dry paper and cleaned up after each time. Keep checking and re-working until the cam turns freely.

The 18V engines have simplex (single row) timing chains, which are fine for normal road use. For high performance use it is best to change to the duplex (double row) set-up. This means changing the crank and camshaft sprockets. The cam gear lock tab and tensioner assembly is the same as for the simplex.

The camshaft endfloat will also need checking, and here I disagree with the manual which says 3 to 7 thou/0.076mm to 0.18mm (possibly mixed up with the crank endfloat, for which the book says 2 to 3 thou/0.005mm to 0.076mm). I set it at around 1 thou/0.0254mm. This will probably mean getting a new keeper

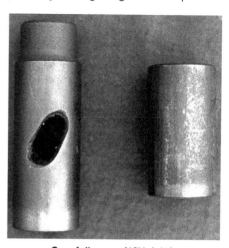

Cam followers (18V right).

plate. The clearance can be fine tuned by sanding the keeper plate, as per the oil pump.

With the endfloat sorted, install the cam (oiled) and fit the cam drive gear to check it with a dial type indicator for run out. The gear will need a slight skim to rectify. Some of the packing washers that go behind the crankshaft timing gear will need to be removed, to compensate for the reduction in thickness of the cam gear to ensure correct alignment of the timing gears. Check this during engine assembly.

As a finishing touch, you may wish to very lightly deburr the edges of the gear teeth with an abrasive cartridge, or flap wheel, and a drill. The timing chain's progress around the gear is made just that little bit smoother as a result, which helps longevity and cuts down power loss.

A badly worn or damaged distributor drive skew gear on the camshaft is a major cause of erratic ignition timing. Unfortunately, the only solution is to fit a new cam. Both the cam and distributor drive gears benefit from a deburr and fettle with a fine file to remove any imperfections that may upset their operation.

Changing from the heavy, original (non-18V) 'milkchurn' cam followers and short pushrods to the 18V shorter followers and long pushrods, gives a small saving in reciprocating weight, and a large saving in money (new late followers and long pushrods work out cheaper than buying the early followers alone!). The valve train geometry will need careful checking however (see cylinder head chapter).

BEARINGS

Contrary to popular belief, bearings themselves rarely fail: the problems are more likely to be a result of dirt, lack of lubrication or incorrect installation,

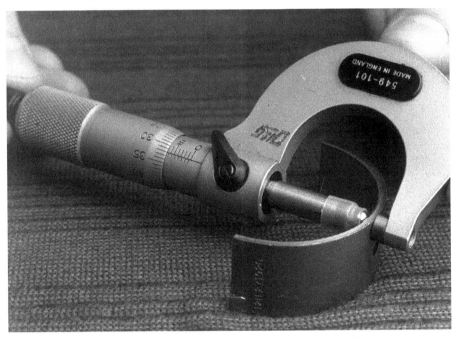

Measuring bearing thickness.

meaning bearing housings or clearances out of tolerance.

Incorrect installation is remedied by the measuring and machining methods given. Lubrication is down to the oil pump – covered later – and correct operating clearances.

You can arrive at the optimum bearing clearance by selectively fitting the bearing shells to individual crank journals. Measure all the shells with a micrometer, get a ball end adapter or use a ball bearing. They don't vary by much, but any slight difference may come in handy. Write each bearing thickness on the back with an indelible pen.

Assemble the con rods and mains with the thickest bearings in the upper half of the rod or main bearing, and torque to spec. Measure the inside diameters with an inside micrometer or snap gauges and note each dimension. The newly machined (or cleaned up) crankshaft journals should then be

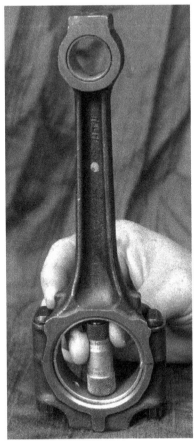

Measuring bearing inside diameters.

carefully measured and recorded. Taking one from the other gives the bearing clearance. You can mix and match the different thicknesses of bearing shells to arrive at the optimum clearance (keeping the thicker shells in the upper halves). If you are a complete perfectionist this may mean purchasing several sets of bearings and checking until all clearances are spot on.

The quick and simple alternative is to use the Perfect Circle product – Plastigauge™. This is a deformable plastic 'string,' supplied in differing thicknesses depending on the clearance to be measured. It is placed across the bearing width and is crushed in place by torquing the bearing cap (with bearing), being careful not to rotate the crank at any time. When the cap is removed the crushed Plastigauge™ can be compared to the scale on the wrapper to give the clearance.

THE OIL PUMP

The standard oil pump is perfectly up to the job of supplying the engine's 'life blood' provided it is in good condition. It should be stripped, thoroughly cleaned and checked, regardless of age or usage, and renewed if the inside of the aluminium casing is scored in any way.

While cleaning the relief valve, give it a tap to check if there is a small 100 thou (2.54mm) spacer wedged in there. This was used on some earlier engines to pre-load the spring, improving oil pressure significantly. One may have been put inside the relief valve or situated behind the nut that keeps the relief spring in the block. Replacement shims are no longer available so don't lose it. It's a good idea to check all blocks and valves, as a spacer may have been fitted if the pump was reconditioned previously.

The problem with the pump is that the rotor tolerances are usually wrong,

Checking oil pump rotor clearance.

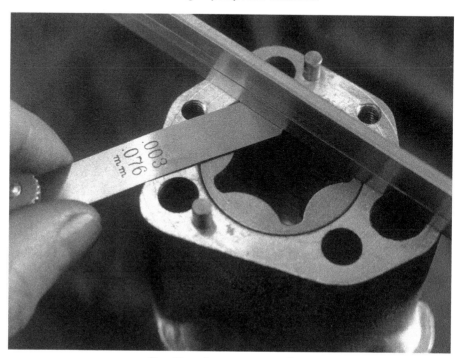

Checking oil pump rotor endfloat.

Oil pump rotors.

Pressure relief valve and spring.

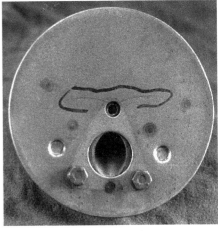

Cracked oil strainer.

even with reconditioned or brand new pumps! Details of how to check the clearances are in the workshop manual. Incorrect clearances for the rotor lobes can be remedied with new rotors. If the endfloat of the rotors is wrong (the most common fault) use fine – 600 to 800 grit – wet and dry paper on a glassplate to reduce either the housing face or the gear height. You can also true-up the cover plate the same way.

Deburr all edges of both rotors with an engineer's stone or a fine needle file, even if brand new.

There is no need to go to the extent of double porting the pump or any of the other modifications given in the old Special Tuning manuals. All that is required is to smooth, deburr and blend the oilways in the pump body and delivery pipe to remove any nasty steps or sharp edges and so ease the oil's passage through the pump.

Either re-use the original pressure relief spring and valve or replace it if in doubt. Always buy Rover original equipment replacements, as some aftermarket valves are plain mild steel and not chromed steel (a satin finish) and can gall in service and stick open. As plated originals are becoming hard to find we substitute a 9/16th ball bearing and use a full length relief spring. The relief spring should have a free length of 3 inches (75mm). Replacements can vary greatly so a shim or small nut may have to be fitted behind it to compensate for any shortfall.

Smooth and blend inlet oilway (indicated) in pump body.

Smooth and blend oilway at position shown.

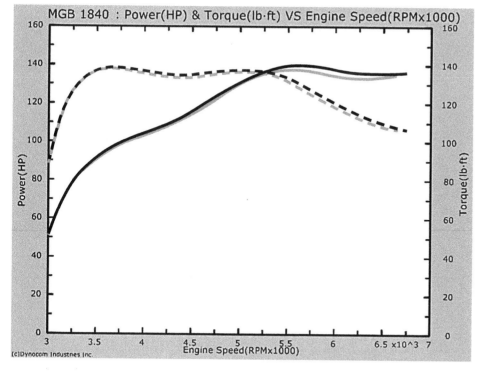

This graph shows the power and torque from the same engine tested with two different oil pressures. We tend to build the engines with 9/16th inch ball bearing relief valves and use full length oil pressure relief valve springs. The lower torque and power curves (pale lines) are with the oil pressure at 100psi. Dropping this to 70psi liberated 3 horsepower at higher rpm. Well worth having on the track!

the relief spring can be shimmed (100 thou/2.54mm) to give 70psi hot.

The only other thing to be aware of is that the back of the oil strainer can crack. This is not really a problem for road use, but, when racing, any oil surge during hard cornering may uncover it sufficiently for air to be sucked up, and that's not good for the bearings. Peace of mind can be restored by welding up the crack, but be careful as the metal is very thin and is easily burnt through, rendering the screen useless.

SUMP (OIL PAN)

For competition the only modification necessary to the sump is to install a baffle in the deep well, similar to the one shown in the Special Tuning books. The only suggested variation is to make it removable for ease of cleaning the sump well. Weld some small angle brackets to the inside of the sump, fit Avdel threaded inserts (Nutserts) and attach the baffle plate (aluminium saves a little weight) with some small bolts and Loctite them in.

Make sure the plate is fitted just above the maximum oil level in the well and that there is a hole to allow the dipstick to reach the bottom.

Gasket sets come with two for the oil pump. The one on the left is for three bearing engines; the other for five bearing engines. Be sure to fit the correct one!

The final touch before reassembly is to lap the relief valve onto its seat in the engine block with a fine paste. This gives a far better seal and improves oil pressure at idle. Thoroughly clean the valve and block passage afterwards, removing all traces of grinding paste, otherwise it will destroy the engine very rapidly – guaranteed!

The engine should then have oil pressure of around 60 to 70psi (hot) when running. Any more is unnecessary and only increases power losses through driving the pump. For race use

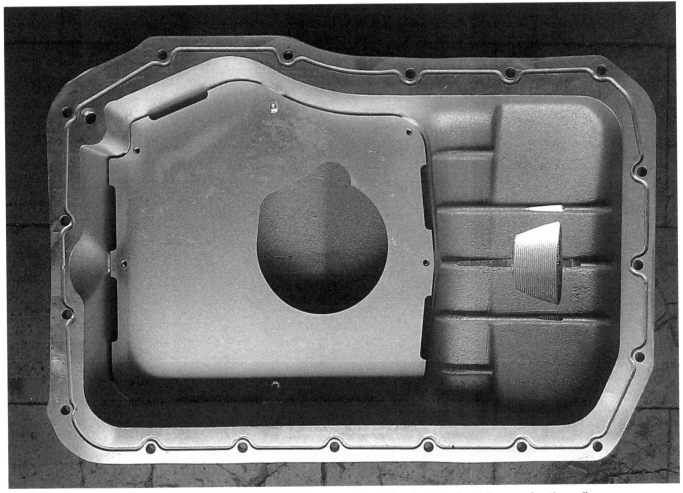

This cast aluminium sump (oil pan) has a removable baffle plate fitted horizontally, covering the well.

GASKETS

The excellent rebuild gasket sets by Payen are available from motor factors and engine reconditioners. The latest B series head gaskets are resin impregnated (very dark in appearance) which helps overcome the B's common problem of a weeping head gasket causing water leaks down the front of the block. Both head and block surfaces must be spotlessly clean (and free from old gasket material if it's only a head swap) for it to work properly. When the engine comes up to operating temperature the resin softens and conforms to the various surfaces which,

combined with a fly cut finish, promotes better sealing.

If you cannot obtain these gaskets, the ordinary composite variety benefits from a thin smear of sealer such as Blue Hylomar or Wellseal to reduce the chance of leakage or blown gaskets. This is particularly important for race applications.

The final variation is the crankshaft rear main oil seal. An uprated version from Rover is available, part number LUF10002 (Sherpa van). This has a superior sealing lip design which goes a long way toward rectifying the weeping rear oil seal problem that affects so many Bs.

Note if the crank has a slight groove in it, caused through contact with the rear seal lip, always fit the new one so that it runs in that groove as well.

FINAL ASSEMBLY

Before final assembly everything will need to be thoroughly cleaned again – bar the block which should have been stored in a clean, safe place. Everything can be assembled as per the manual, using the engine oil you intend to run as a lubricant for the bearings and pistons.

The last remaining thing to do is check the crankshaft endfloat. This is governed by crescent-shaped thrust

Flywheel runout is checked and corrected as necessary.

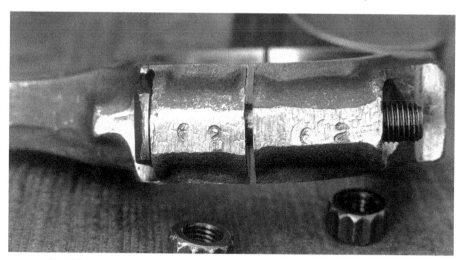

Don't forget to reassemble connecting rod halves the correct way around.

(0.076mm to 0.102mm) endfloat (nearer 4) rather than the 2 to 3 thou (0.005mm to 0.076mm) specified in the manual.

If the clearance is too large try another – oversize – thrust bearing. If the clearance is too little, increase by lightly sanding the thrust surface on fine 600 grit wet and dry on a glassplate or other flat surface – work slowly and measure frequently.

When it finally all comes together you should have an engine that turns over quite freely by hand. All that's then required is to fit it in the car and run (break) it in.

STARTING THE NEW ENGINE

With a brand new engine it is best to use a specifically formulated running-in oil or cheap 20/50 oil for the for the first start and the initial running-in period. Modern oils are excellent at preventing engine wear, unfortunately! If the engine is run in using a modern oil containing all those high-tech additives, there is a good chance it will burn oil for the rest of its life! Those additives prevent the rings and bores bedding in properly. In this instance, you need the engine to wear in a controlled manner so the tiny surface ridges (or asperities) left in the bores by the machining processes flatten rather than bend and work harden. It is the bending deformation that forms tiny pockets that retain oil, which subsequently burns during combustion. Running in oil is usually straight 30 weight oil without additives. Fill the engine to the maximum mark on the dipstick. Moderate overfilling is acceptable to allow for the oil cooler and filter.

Do as much as possible to ensure the engine will fire first time: set the timing, check there is fuel to the carburettor(s), ensure the battery/ batteries are fully charged, etc. Remove

washers situated each side of the centre main bearing. Sometimes the machine shop will clean these faces so it's a good idea to check.

Install the crank with the bearings in place and lightly nip up the main bearing caps. Then gently tap the crank to and

fro with a soft-faced mallet to seat the bearings and thrusts. Torque the bearing caps to spec. Fit a dial gauge to the end of the crank, measuring along its axis. Pry the crank back and forth with a large screwdriver and read the endfloat from the gauge. I prefer having 3 to 4 thou

the rocker cover and, with the live feed to the coil disconnected and the sparkplugs removed, spin the engine over on the starter until oil can be seen flowing from each rocker position. This should take about 15 to 20 seconds (30 at the outside). This is a good check to ensure that all is well in terms of oil feed to the top end. Absence of oil, even after extended engine cranking, should be investigated and remedied. Refit the cover and spin the engine over on the starter until the oil light goes out or at least half normal pressure is read on the gauge, which can take some time. Once that has happened re-check the oil level.

Replace the plugs and coil feed wire and start the engine (use of choke is OK but avoid running the engine for a prolonged period with it on), bringing it immediately to around 2000rpm. Run it for 10 to 15 minutes or so at between 2000 and 2700rpm to bed-in the cam (see cam chapter) and allow the oil to circulate fully. Don't leave it idling with the choke out.

While it's running (put a small block or wedge in the throttle linkage to keep the rpm high – or get an assistant) check all around for any leaks or problems.

If something needs investigating or tightening, shut the engine off, adjust, restart and continue for the remainder of the bedding-in period.

After bedding in the camshaft reset the tappet clearances (valve lash) – remember you need a hot clearance. Check and set the ignition timing to the required setting. Set the carburation. Check the oil level and check for oil or water leaks. Once cool, do a spanner check around the engine and ancillaries for any loose nuts and bolts.

RUNNING IN

A short period of considerate use of a new engine gives the various components time to settle and allows the piston rings to bed-in to the fresh bores.

It's just as possible to spoil a new engine by running it in too carefully as it is by thrashing the living daylights out of it. A compromise is needed. If possible, get the first 100 miles or so done in one go. This allows everything to reach operating temperature and warm up thoroughly. Vary the engine speed as much as possible and allow the engine to work reasonably hard. As long as all the gauges read normally the occasional burst of acceleration to 5000rpm or so

in the lower gears is permissible. The gas pressure in the cylinder at these speeds pushes the rings out against the bore walls and greatly helps bedding-in. The short bursts of power will allow the rings a cooling and settling period, in-between the spells of working hard. Avoid prolonged high revs at this stage of the engine's life. The only other thing to avoid is excessive load. Try not to 'lug' the engine at low rpm in a high gear, change down for hills if necessary or when trundling around town.

It is not a good idea to bed-in an engine by spending ages on a motorway or dual carriageway with it running at a constant rpm for a long period. Give it some variety in its work!

After the engine has done around 500 miles, change the oil and filter. Refill with the brand of oil (20/50) of your preference, or as recommended by your engine builder/supplier. Check the tappets (valve lash), cylinder head nuts and ignition timing. You can then start to enjoy your engine's new performance.

Give it about 1000 miles and you can book a rolling road session to have the various settings optimised and unleash the engine's full performance potential.

Chapter 7
Camshafts

The job of the camshaft is to open and close the valves at the right time, in order to fill the cylinders with the fresh mixture before combustion, and empty them afterwards. This sounds fairly straightforward, but how the job is accomplished has a significant effect on torque, horsepower, the operating range of the engine and its driveability. As I brushed over the subject in Chapter 2, a brief description of how the cam works in relation to the four stroke cycle will not go amiss here.

The camshaft provides a means of converting rotary motion into the reciprocating motion that opens and closes the valves. The cam is driven off the crankshaft by the timing chain and always rotates at half the crank speed (rpm). It takes two full rotations of the crank – which is one rotation of the cam – to complete the four stroke cycle. It is very important that the camshaft is installed in proper relationship to the crank so that the valves are opened and closed at the correct time during the piston stroke. Setting this relationship of cam timing to crank timing is what is meant by the term timing, or degreeing, the cam.

Let's make a few generalised observations, beginning with the exhaust stroke. With the piston on its way down the cylinder during the power stroke, the exhaust valve opens before the piston reaches Bottom Dead Centre (BDC), allowing remaining combustion pressure to begin leaving the cylinder (called blowdown). The piston moves back up the cylinder, pushing the remaining exhaust gas out. Meanwhile, as the piston approaches the top of its travel – Top Dead Centre (TDC) – the inlet valve starts to open before the exhaust valve has closed. This takes advantage of a vacuum created by the rapidly departing exhaust gas in order to start drawing the fresh air/fuel mixture into the cylinder. The exhaust valve closes fully just after the piston begins its movement back down the cylinder, continuing the intake stroke. The inlet valve then remains open after the piston passes BDC and begins the compression stroke. This takes advantage of the reluctance of the now rapidly moving incoming air/fuel mix to stop, so it keeps piling into the cylinder past the inlet valve. With both valves shut the piston's motion up the cylinder continues the compression stroke, before combustion is initiated and the whole process starts over.

Regarding camshafts, compromises are needed to achieve the optimum trade-off between low speed torque and high rpm horsepower. During the process of choosing a cam, lift and duration are the most commonly used criteria for determining a camshaft's suitability for a particular application. But nowadays more attention is being paid to a cam's lobe separation angle – also called the lobe displacement angle. This has mostly replaced the previously used term, overlap.

CAMSHAFT TERMINOLOGY AND OPERATING PRINCIPLES

The camshaft lobes are the lumps (technically called eccentrics) responsible for converting the rotation of the cam into the up and down movement of the valves, via the cam followers, pushrods and rocker arms. The actual shape of the lobe is termed the profile, and there are hundreds of possible profiles within a lobe shape that have the exact same lift, opening and closing points.

The base circle is the round portion of cam lobe where the lift is zero and the valve clearance (tappet) adjustments are made. During this time the valve is allowed to remain on its seat, sealing the port from cylinder and also transferring some of its heat to the cylinder head via the seat. The tappet clearance allows for the heat expansion of the valve and valvetrain.

The clearance ramp is a short section that allows transition from the base circle to the point on the profile where measurable lobe lift occurs. The rate of lift here is fairly slow as the shape takes up the various valvetrain clearances gently, ready to begin lifting the valve.

The area of the lobe which does the lifting is the flank (or ramp). This provides positive acceleration of the follower to generate the lift, followed by a region of controlled negative acceleration (slowing down) until the valvetrain pauses fractionally at the cam nose (full lift). The opposite flank then controls the positive acceleration of the valvetrain away from full lift, with a further region of negative acceleration prior to lowering the valve gently back onto its seat before reaching the base circle again. Most lobes are ground with mirror image lift curves on opening and closing sides.

An asymmetric lobe is ground with a different lift curve on the opening and closing sides, the valve may be opened quite rapidly and then closed more slowly, or vice versa.

Lift is the maximum distance that the valve is raised off its seat, given in either thousandths of an inch or millimetres. Theoretical lift (as used in most cam catalogues) is calculated by multiplying the camshaft given lobe lift by the (note!) theoretical rocker arm ratio. Net lift is the actual physical lift of the valve measured at the valve (usually during the blueprinting of the valvetrain assembly). There is always quite a difference between the two. The net lift takes into account the tappet clearances, any rocker arm ratio differences and different rocker pad shapes, as well as pushrod and other mechanical deflections in the valvetrain. Sometimes the cam lobe lift itself is different, for some reason, from the specification given on paper (a dodgy grind?).

Each lobe of the cam is ground with a slight taper across it. This taper, together with the slight crown on the contact face of the follower, compensates for any misalignment of the follower bores and also spins the follower in its bore to reduce the wear between it and the cam, helping to prolong component life.

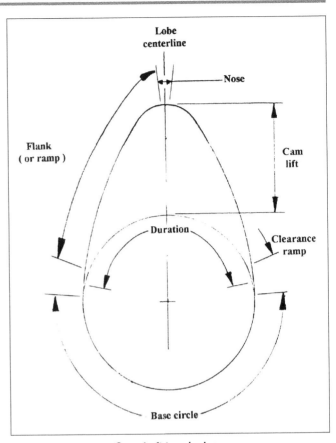

Camshaft terminology.

Rate of lift is how fast the valve is opened and how many inches of lift per degree of crankshaft rotation. It is the diameter of the cam follower base that dictates the maximum rate of lift from a cam profile. While lifting the follower the flank of the cam wipes across the follower's width, from the middle to near the edge and back to the middle. Too high a rate of lift can cause the flank of the cam to dig into the edge of the follower, which is not conducive to long valvetrain life.

Duration is the amount of time the valve is held open, measured in degrees of crankshaft rotation. As advertised, this value varies tremendously depending on where the manufacturer decides the point of valve lift begins.

Lobe separation angle (LSA, also

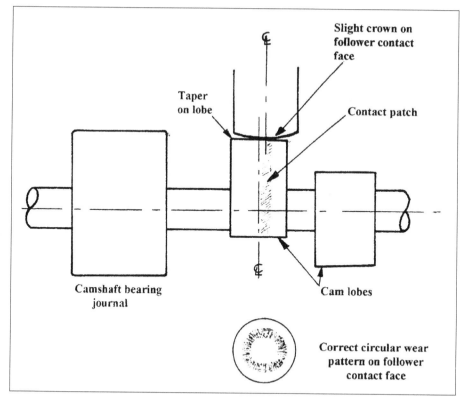

Slight crown on follower contact face

Taper on lobe

Contact patch

Camshaft bearing journal

Cam lobes

Correct circular wear pattern on follower contact face

Cam lobe and follower relationship.

called lobe centreline angle or LCA) is the angle between the centrelines of the inlet and exhaust lobes, given in camshaft degrees. This is fixed when the cam is ground and is becoming more widely used in place of the previously used, less accurate term, overlap.

Overlap is the period of time when the intake and exhaust valves are open simultaneously. As mentioned previously, this happens around the end of the exhaust stroke and the beginning of the intake stroke – as the piston approaches and leaves TDC.

Single pattern cams have an identical lobe profile for both the intake and exhaust lobes. A dual pattern cam has different lobe profiles for the intake and exhaust lobes, usually to overcome some deficiency in the performance of a particular engine's intake or, more commonly, exhaust system.

Lastly, there is timing, or degreeing the cam. Timing the cam does not change its duration, lift or lobe separation angle, as these are all fixed when the cam is ground. It merely allows the point at where the valves are opened relative to piston stroke to be varied. Moving the cam alters the rpm at which peak torque and horsepower occur. Retarding the cam (opening the valves later in the cycle) moves peak power to a higher rpm and can increase horsepower but at the expense of low rpm torque. Advancing the cam (opening the valves sooner in the cycle) has the opposite effect, moving power to a lower rpm and increasing low rpm torque. These effects are not guaranteed, however, and do vary depending on the particular engine and camshaft being used, so experimentation is the only way to find out for sure.

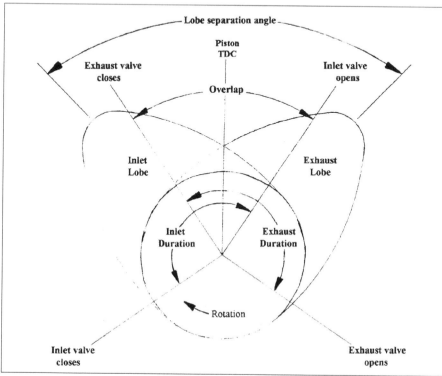

Lobe separation angle

Piston TDC

Exhaust valve closes

Inlet valve opens

Overlap

Inlet Lobe

Exhaust Lobe

Inlet Duration

Exhaust Duration

Rotation

Inlet valve closes

Exhaust valve opens

Inlet and exhaust lobe relationship.

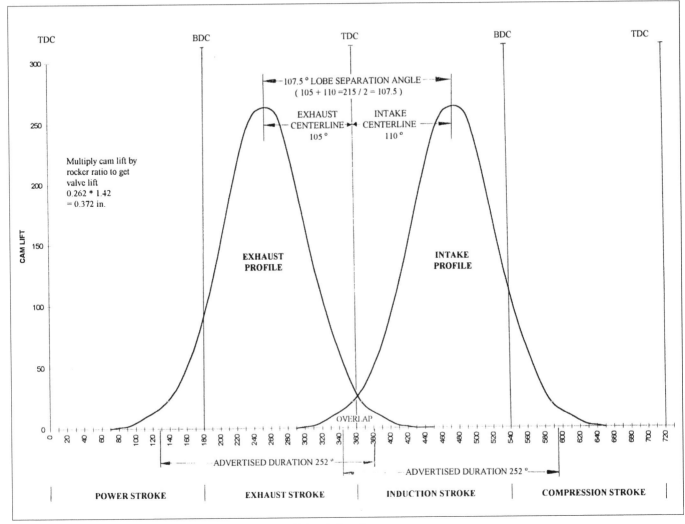

Valve timing diagram.

Advertised duration can be given as including the lobe clearance ramps, where no valve lift at all is taking place, or given as the duration at 0.050 inch (1.27mm) cam lift – as used by American cam companies – with many permutations in-between. All these different methods of measurement, coupled with the hundreds of different profiles and rates of acceleration available within a lobe shape, mean this is a fairly pointless method of trying to compare cam specifications. This is where lobe centreline angles come in (calculations in

Appendix). The lobe separation angle can remain constant when the duration or rate of lift (profile) is altered, while the overlap doesn't. So, in effect, two cams with the same lobe separation angle can have dramatically different overlap, depending on their duration.

For a given profile, then, as lobe separation is widened (say, from 102 to 107 degrees) overlap decreases, giving increased cylinder pressures and more torque. This gives a smooth tickover and improved low- to mid-range torque but limits high rpm power. As the lobe

separation is tightened (say, from 110 to 106 degrees) the overlap increases and cylinder pressures at low rpms are reduced, as is torque and horsepower. This gives rise to a lumpy tickover but better mid-range torque and high rpm power.

If you intend to use high ratio rockers with anything other than the standard cam, look for a profile with between one and three degrees more LSA than standard; from the original 107 to, say, 110 degrees. This will counter the affect the higher rocker ratio has of closing up

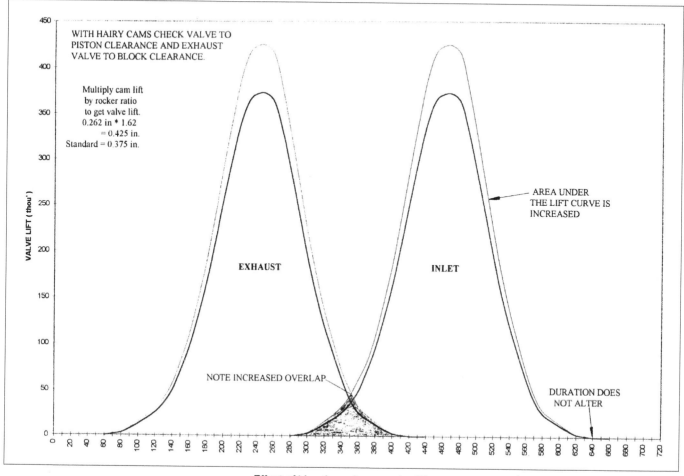

WITH HAIRY CAMS CHECK VALVE TO PISTON CLEARANCE AND EXHAUST VALVE TO BLOCK CLEARANCE.

Multiply cam lift by rocker ratio to get valve lift.
0.262 in * 1.62 = 0.425 in.
Standard = 0.375 in.

AREA UNDER THE LIFT CURVE IS INCREASED

EXHAUST

INLET

NOTE INCREASED OVERLAP

DURATION DOES NOT ALTER

Effect of hi-ratio rocker on valve lift.

the apparent LSA, due to the increased valve open area (more overlap at TDC). Otherwise the engine will become more 'cammy' than would ordinarily be the case. An additional few degrees of lobe separation can be specified with a new from blank cam purchase. Most cam companies are willing to grind one of their profiles with some extra spread to suit high lift rockers.

Scatter pattern cams used different full lift timing and lobe separation angles for each cylinder of the siamese pair in an effort to overcome the flow interference problems the siamesed inlet and exhaust ports present. Going to this extent with the cam is probably best saved for race applications, where every last ounce of horsepower is wanted.

Most MGB tuned cams are very good at sacrificing bottom end power and not gaining top end horsepower, only increasing the engine's revability. Quotes of horsepower increases can adorn camshaft sales blurb, but you need to be very careful where the gains actually occur. While a cam may allow the engine to rev more and give more power, there may not be any gain at the 5000rpm, or so, point where peak power occurs, but it makes more power at rpm beyond this point. Should this be considered a power gain or an extended powerband? Even those cams that purport to be replica BLMC Special Tuning profiles do not seem as effective as the original factory tuned cams (now virtually unobtainable).

Don't dismiss the standard MGB cam, it's a superb design that is currently unequalled in ability to make good horsepower yet still retain excellent tractability (see Power Recipes in Chapter 4); an important consideration for enjoyable motoring.

Without going into further long-winded explanations of camshaft characteristics, the best advice I can give is to describe the performance of camshafts experienced over the years.

This, combined with the camshaft tables (Piper and Kent profiles) in Appendix I and the power recipes in Chapter 4, should help you choose a cam for your needs.

Piper HR270 (Magnum). An excellent, all-round performance cam for all MGB engines. There is no loss in flexibility (pulls strongly from around 2000rpm) with a slight gain in power and revability. The HR270/2 and BBP270 profiles do not give any real gains over the original Magnum.

Piper HR285 (Magnum). An excellent fast road cam, giving a lumpy tickover and slight loss of low rpm power. It gives good improvements in mid-range and upper rpm power and extended revability. At the wheels can give bhps in the high 90s with modified engine and twin 1¾ SUs. The BBP285 is marginally more tractable but, again, no real improvement over the original.

Piper BBP300. A good powerful rally cam, but won't pull foot flat acceleration below 3000rpm.

Piper HR320. A very powerful race cam – with a full race head it intensely dislikes working below 4000rpm, but has very strong performance above that.

Piper BBP320. A very strong cam, as above.

Kent & Piper 714. A mild road cam that's only good for very high revs without delivering any real power gains and which really does not work well for road use – it doesn't seem to deliver anything except a lumpy idle. It is a copy of the BLMC Special Tuning profile, designated 731 when used in A-series engines. We understand from a friend who competed in minis in the '60's that it was designed as a rally cam for events on loose surfaces such as shale or gravel where revs were needed combined with a gentle power delivery; no abrupt increase in torque that would compromise traction.

Kent 715. Works best with sidedraught Weber carburettors rather than SUs – bhps of mid- to high 90s with modified head and Weber 45, but only low 80s with SUs.

Kent 717. Adds a couple of horsepower at peak, loses a couple at 3000rpm when used with modified head. It is very good from 4000rpm upwards, with good torque from 3500-5500rpm. An excellent, long-legged motorway cam.

Kent 718. A very hairy cam with nothing below 2600rpm – a bit intractable.

Kent 719. Not for road use at all. If you're up at this level it's better to go for the 719 scatter pattern, giving amongst the highest power outputs and torque spread for a full race MGB engine. Scatter pattern cams are only necessary for rally and race use, where the scatter design brings them 'on the cam' some 500rpm earlier than would otherwise be the case. For this type of use it is best to have the cam ground from a new blank and cross-drilled to improve oil feed.

VALVE SPRINGS

The valve springs provide a positive means of closing the valve and keeping the follower in contact with the camshaft so that valvetrain motion is controlled. To perform correctly the valve spring must be matched to the camshaft design and the engine's peak rpm level. They should have sufficient pressure to keep the follower in contact with the cam lobe and to ensure the valves don't float (momentum causes the valve to hang open when the rest of the valvetrain should be heading down) or remain open at the wrong time. But they should not have so much pressure that they rob horsepower or wear out the cam. It takes force to compress the spring – multiplied by eight springs – force that resists the cam's rotation, frictional losses, remember.

Stiff valve springs can also cause the rocker assembly to flex, which puts a strain on the components and alters the valve timing, so matching valve springs to engine is critical.

The standard factory single springs (18V engines) are suitable for use with cams giving up to 400 thou valve lift, and allow 6200rpm before valve float imposes a physical rev limit. The factory doubles allow 430 thou valve lift and 6400rpm before valve float.

For cams with more lift and greater rpm ceiling, typically competition usage, you can opt for the cam supplier's recommended spring (they know best) or the stronger springs available from the various MG specialists. If high ratio rockers are to be used with a high lift cam, then the Moss low coil variety is essential to avoid coil bind.

BEDDING IN THE CAM

The first moments of operation of a new cam are vitally important with respect to whether or not it will live long. To help it survive, coat the cam lobes and bottom of the followers with a graphite based high pressure lubricant such as Graphogen, although a molybdenum assembly grease is equally good.

Do as much as possible to ensure the engine starts fairly quickly (set ignition timing etc) and, once running, immediately take it to around 2500rpm and keep it there for about 15 to 20 minutes. This high rpm increases the oil supply, reduces load on the cam and gives the lobes and followers time to bed-in. We usually rev the engine gently at between 2000 and 2700rpm during this time, just to vary the speed and load. Do not let the engine idle at any time during this period. If you need to do something to the car, switch off, adjust, then take the engine speed back up for the remainder of the run-in period.

Chapter 8
Carburation

In an internal combustion engine, the piston moving down the bore creates a vacuum in the cylinder and inlet manifold when the inlet valve is open. The air outside the engine, being at a higher (atmospheric) pressure than the inlet tract, moves in to fill the depression and, on its way, passes through the carburettor where the fuel is added to enable the engine to produce power.

The main objective of any carburettor is to supply a mixture of fuel and air to the engine in a form which can be rapidly and completely burnt. To achieve this, the air and fuel mix needs to be supplied in vapour form. Therefore, the carburettor has to be able to break up the liquid fuel, or atomise it, as well as disperse it effectively into the air passing into the engine. How well the carburettor accomplishes this considerably affects engine combustion efficiency and, hence, how well the engine performs. To top it off, the carburettor must also be able to supply varying amounts of

fuel and air to cope with the engine's varying speeds, and dispense different amounts of fuel to the air depending on the engine's power requirements.

The theoretical ideal amount of air to fuel for complete combustion is called the stoichiometric ratio, which is 14.7 parts of air to 1 part of fuel (by weight). But this does not give either maximum power or minimum fuel consumption from an engine. Maximum power is usually generated with an air/fuel (A/F) ratio of 12.5:1 (less air), while a comfortable cruising condition for the engine is around 13.4:1. More modern engines run using different A/F ratios to these theoretical ideals, as manufacturers strive for more power, more economy, and fewer emissions. However, though these figures are valid for the majority of older cars, they are by no means gospel, every engine has unique requirements for best performance.

It's possible to determine the

amounts of fuel being delivered to the engine by using an exhaust gas analyser to measure the amount of carbon monoxide (CO) being produced. CO is a by-product of the combustion process and there is a known relationship between the amount of CO present in the exhaust gas and the air/fuel ratio. Nowadays, air fuel ratio meters are available that use lambda sensor technology to measure the A/F ratio directly from the exhaust, which can be converted back to %CO if necessary. Combining an analyser or lambda A/F sensor with a rolling road, to simulate driving conditions, allows the mixture supplied by the carburettor(s) to be checked and adjusted if necessary, to see if it is giving the optimum amount of fuel for best engine performance. Best power is with a CO reading of around 5 per cent, and best economy with a reading of around 0.5 to 1 per cent. For cruising, a reading of around 2 to 3 per cent CO is a happy medium.

Too rich a mixture will cause carbon to build up on the combustion chamber and piston crown, as well as interfere with sparkplug performance. The excess fuel also contaminates the engine oil, impairing its performance and causing premature cylinder bore wear by washing off the thin film of lubricating oil between the bore and the piston.

On the other hand, too lean a mixture can cause the engine to overheat and may result in burnt valves and damage to the tops of the pistons. It is also more difficult to set alight, so highlighting any inadequacies in the ignition system as well as losing engine performance.

An uprated fuel supply system, comprising pump, filter and pressure regulator will be needed if the engine is going to be producing much more power than standard – more detail later in the chapter.

A good quality throttle linkage is also a must for ease of use and setting-up, together with a low restriction air filter – again, more about this later in the chapter.

I'll cover the two main types of carburettor most commonly used on a B which are, firstly, the SU as fitted as standard to the majority of cars and, secondly, the sidedraught Weber or Dellorto carburettors usually associated with a performance upgrade.

SU CARBURETTORS
The SU has to be one of simplest yet most effective carburettors fitted to production vehicles. Its superb design provides a long lasting, robust and mostly trouble free means of fuelling an engine. But because of its commonplace usage it has never had the high performance image it deserves.

The SU carburettor is easily capable of providing power outputs from the B series only very marginally surpassed at very high rpm by competition-type sidedraught installations. Any minor problems encountered with SUs are usually easily rectified by a little straightforward maintenance as described in Chapter 14.

SU carburettors work on the constant vacuum principle (also called constant depression). Inside them a sliding piston completely blocks the air passage when the carburettor is not working, resting at the bottom of its travel by the influence of gravity and help from a light spring. The chamber the piston works in is effectively air-tight, air entering or leaving through a hole in the base of the piston only which is positioned on the downstream (vacuum) side.

On starting the engine, as the starter turns it over, the piston's obstruction of the airflow into the engine creates a vacuum on the downstream side. This vacuum is transferred to the piston chamber through the hole. As the pressure underneath the piston is around atmospheric, the difference in pressure above and below the piston lifts it from its rest position. The amount the piston is lifted depends upon the amount of airflow into the engine, giving what is, in effect, a self-adjusting carburettor.

A tapered needle is attached to the piston, which fits into a single circular jet beneath it. The fuel is issued from the space around the diameter of the needle as it fits in the jet. Since the designers can calculate the amount of piston lift according to amount of airflow, the diameter and the taper of the needle can be chosen in order to block a certain amount of the jet so that only a certain amount of fuel is allowed through, giving the correct mixture at all engine speeds.

At tickover there is enough suction from the engine to pull fuel from the jet, so no idle circuit is needed. There is also no separate acceleration circuit as, again, the carburettor design allows for the extra fuel necessary for acceleration to be supplied from the single jet. The central rod that guides the piston is a tube, which is filled with oil. There is a loose fitting brass bush mounted on a central spindle that sits in the oil, so when engine airflow demand increases suddenly due to acceleration and the piston tries to lift, it is stopped temporarily by the hydraulic action of the bush in the oil. The temporarily increased vacuum created by the blockage from the piston draws additional fuel from the jet, richening the mixture.

It is interesting to note that a correctly set-up SU has more 'fuelling sites' than computer controlled fuel injection systems, ie more ability to specifically alter fuelling to suit varying load and throttle positions, – not bad for a 'prehistoric' design!

Modifications
The normal carburettor throttle butterfly is a brass disc which is attached to a spindle connected to the accelerator pedal via a linkage. On some SU carburettors, the butterfly is fitted with a spring-loaded valve, and the idea behind this is to reduce engine emissions during periods of engine overrun. With the engine at high speed, if the throttle is closed suddenly the enormous increase in vacuum immediately draws any fuel that has been stuck on the walls of the inlet manifold into the cylinders, causing a short period of extreme richness and high exhaust emissions. The valve is meant to remedy this by opening during these extremely high vacuum conditions and allowing air through, reducing the vacuum and allowing the mixture to burn better with the extra air. This causes a noticeable lack of engine braking in cars that have these valves in the carburettors. The problem is the springs holding the valves shut get tired with age, resulting in the valves leaking when

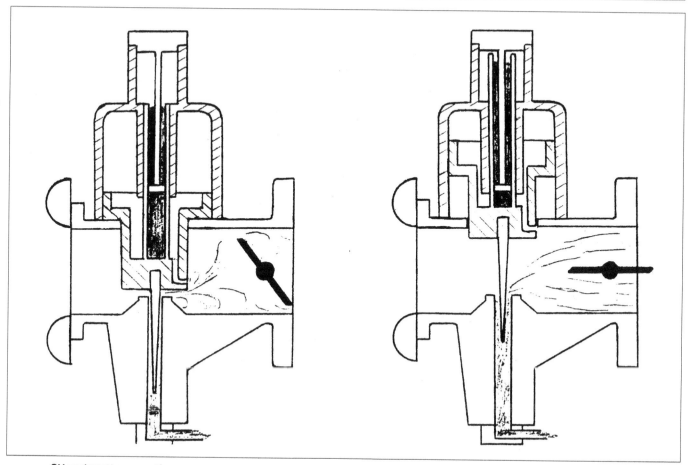

SU carburettor operation.

the engine is at tickover, which gives an uncontrollably high idle speed.

The valves also present a large physical obstruction to airflow, so it's a good idea from the point of view of power gain and improved idle to change the butterflies on carburettors so equipped to plain, non-valved ones. Failing this, a temporary solution to the idle speed problem is to solder the valves so they cannot operate.

On the subject of carburettor butterflies, I can mention a few modifications to increase airflow capability as these come within the range of competent DIY skills. If you take a look at the throttle butterfly in the carb, you will see it is held onto the

SU throttle butterflies: plain and late poppet valve type.

A view through the carburettor showing restriction caused by spindle, screws and poppet valve.

one half of the slotted spindle entirely, leaving a shaft with a flat on which to mount the butterfly. The countersunk retaining screws should be replaced with round head machine screws and, again, secured with threadlock.

It isn't necessary to modify the shape of the SU throttle butterfly at all. Thinning it can cause sealing problems if you're not careful and, if taken to extremes, can cause it to distort in use.

It can be useful to smooth and deburr the inside of the carb's airflow passage, removing any sharp or protruding edges and blending in any changes of shape, but leave the bridge at the bottom of the carb – where the jet is – untouched.

A standard 1.5 inch (38mm) SU flows 146cfm on my flowbench, radical modification of the butterfly spindle gained 10 per cent more flow, while smoothing and blending of the flow passage added a further 3 per cent, resulting in a flow of around 165cfm. A standard 1.75 inch (44.5mm) SU flows around 204cfm and responds in a similar positive manner to the modifications.

throttle spindle by two small countersunk screws, which have their ends spilt and spread to stop them falling out. The ends of these screws stick out into the airstream quite a bit, and tests on the flowbench have shown that filing them flush increases airflow a bit. The screws will then need securing in place with a spot of thread locking compound.

Further airflow gains can be had by filing the spindle where it holds the butterfly evenly on both sides, to reduce its width and create a more streamlined shape. The final and most drastic modification involves removing

Standard and modified butterfly spindles (modified reduced to 40 thou 1mm aerofoil section between screws).

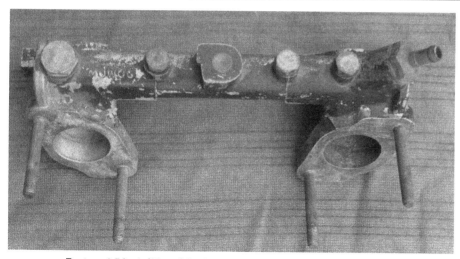

Factory 1.5 inch (38mm) SU inlet manifold (straight shot into ports).

'tripped' airflow to lift fuel from the jet and help disperse the fuel droplets. Reshaping seems a good idea in theory, but if the factory hasn't altered this region in all subsequent redesigns the message is leave well alone!

The factory inlet manifold for the SU is an excellent piece of equipment which can be utilised through all states of engine tune. If the cylinder head ports have been modified and opened up at the manifold face, the manifold itself will benefit from being opened out to match the port sizes, so removing any obvious step there may be. A carbide burr works best on the aluminium, dipped in (or sprayed with) a little WD40 occasionally to reduce clogging. The same applies if a larger 1.75 inch (44.5mm) SU is fitted; just remove any step there may be between the carb and manifold. Don't be tempted to block or obstruct the balance passage in the inlet manifold that links the two carburettors together, as it's there to help reduce the pressure fluctuations in the manifold caused by the valves opening and closing, and so smooth carburettor operation. Flow

The switch to a larger 1.75 inch (44.5mm) SU is only required for very fast road large bore engines with hairy cams, competition or race use. The larger air passages lose out slightly in making power in the sub-2500rpm region while, compared to the 1.5 inch (38mm) SU, performance gain is only possible at above 6000rpm, when the bigger carburettor allows more airflow.

There are no other modifications

I've tried that increase airflow without actually compromising the operation of the carburettor.

In no way do I condone the idea or practice of filing or reshaping the bridge at the bottom of the carburettor where the jet is located in order to increase the flow. The designers experimented with all shape permutations possible and settled on this flat and square shape for a reason, possibly using the deliberately

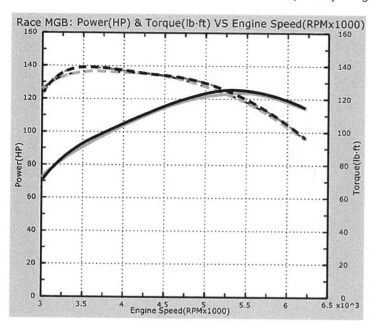

Race MGB: Power(HP) & Torque(lb·ft) VS Engine Speed(RPMx1000)

The graph shows the bhp and torque difference between HS4 and HIF4 carburettors.

The carburettors were tested on a limited modifications regulations race series engine: 1867 (+60) block, 1.625 inch inlet valves and 1.343 inch ex valves, fully modified head, unlimited cam choice, unlimited CR, 11/2 inch SU carburettors. The darker dotted line and solid line are the HIF4 power figures. With the HS4 SU's the max wheel output was 123.7bhp (146bhp at the engine) and 136.5lb/ft of torque. Max power using the HIF4 SU carburettors was 126.7bhp (149bhp at the engine) and torque was 139.1lb/ft. There is little to choose between them for road use. The carburettors tested were in very good condition; worn carburettors would not have produced as much bhp or torque, so differences would not have been as marked. For race use the HIF4s offer a power and torque advantage, plus, with the integral float underneath the fuel supply jet, there is less likelihood of suffering fuel surge under aggressive acceleration or deceleration conditions.

testing showed a 1cfm gain when the hole was blocked and smoothed over, but the engine lost 2 horsepower and ran very raggedly on the rolling road.

SIDEDRAUGHT CARBURETTORS

The sidedraught carburettors from Weber and Dellorto work on the same venturi (choke) principle as the SU, but, whereas the motion of the piston in the airflow passage creates a variable (area) venturi with the SU, creating a constant depression (vacuum) at the jet for different engine speeds, the sidedraughts use a fixed area venturi which creates a variable depression (vacuum) in the carburettor with varying engine speeds.

A typical twin choke sidedraught has two intake barrels, each containing a venturi or choke (hence the name) that sit each side of a common fuel reservoir (float chamber). The two throttle butterflies are connected by a common shaft to act as one unit. As the air is drawn into the carburettor its speed increases as it passes through the venturi, the highest air speed being at the narrowest section of the venturi. As the air speed is increased the air pressure falls below atmospheric. A fuel outlet is located in the venturi at its narrowest section and the low pressure draws the fuel out from the float chamber. By mixing a little air with the fuel before it is drawn out of the jet, the fuel droplet size is reduced and better fuel atomisation is achieved, using what are called air correctors. The only problem is that as the air speed increases due to increased demand from the engine, the amount of fuel drawn out increases by a disproportional amount and the mixture becomes rich. Conversely, when engine speed is decreased the mixture becomes weak. To overcome this, and the other problems of idling, acceleration

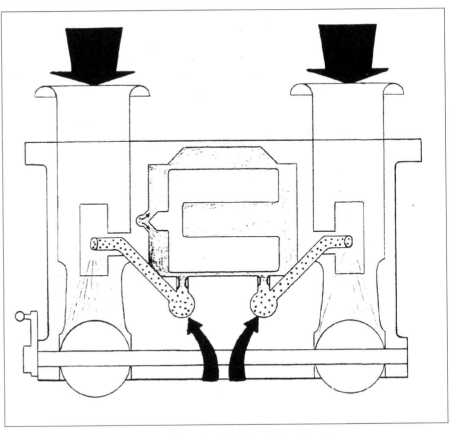

Schematic of a typical sidedraught carburettor.

enrichment and cold starting, the carburettors use a variety of fuel supply jets and internal air and fuel control passages in order to deliver the correct fuelling at all engine speeds and loads.

The chokes can be changed for ones of different sizes, allowing the carb to be matched to different engine applications. With all the other changeable components such as idle jets, main jets emulsion tubes, etc, it means these carburettors can be calibrated accurately to supply the correct fuelling for the wide variety of engine applications. All of these changeable internal parts are available off the shelf from a carburettor specialist.

The intake barrels for these carbs come in a wide variety of sizes: 40, 45, 48, and – Weber only – 50, and these give the carbs their names. For the majority of performance road usage on a B, the single carb should be at least a 45, with the 48s for rally use and even up the 50 for race use.

These carburettors are superbly made precision instruments which have been around long enough for a great deal of calibration information and knowledge to have been amassed about them. This information allows the Weber or Dellorto experts to supply, from experience, the necessary calibration parts to suit a particular application, so a carburettor can be ready to install and use straight from the box. However, while the settings will be somewhere near what is required they will not be tailored to suit your engine specifically. For the fine tuning operation a rolling

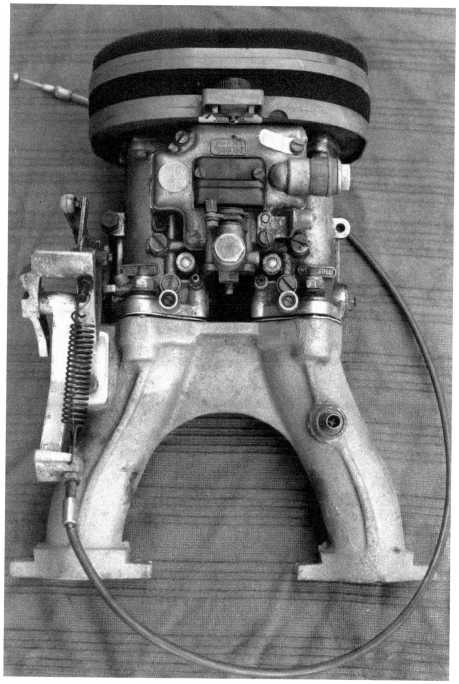

Sidedraught carburettor with long inlet manifold.

There are a couple of long manifolds available that will require the inner wing panel to be adjusted with a large soft hammer to accomplish this, though whether you wish to go quite this far is up to you. The short manifolds tend to have tight turns before they meet the head, and this can cause a slight bias of the mixture flow toward the outer pair of cylinders – which run rich while the inner cylinders go lean. It is the reluctance of the heavier-than-air fuel droplets to change direction rapidly that causes this problem. The longer inlet manifolds (not the ones with the straight runners but those with a curved shape) allow the manifold a much straighter approach to the head, so the fuel is presented in a more unbiased manner to both cylinders. If you have any doubts about whether a manifold does fit easily, ask the supplier.

If you are going to retain the vacuum advance facility on the distributor (a wise move for improved part throttle performance and economy), check the carburettor has provision for a vacuum take-off point. This should be situated on the upstream side of, and very close to, the butterfly, so there will be no vacuum signal until the throttle is opened slightly and the tapping is exposed to the airflow. Having the take-off on the engine side of the throttle butterfly – so the distributor receives a vacuum signal at all times – will create the wrong advance characteristics for the engine.

If the carburettor does not have a vacuum take-off facility you will have to drill and tap the manifold on the downstream side of the butterflies, on the top of both the limbs, or on the carb mating flange of the manifold, and install a couple of small fittings to connect the vacuum take-off pipe to. The alternative is to Araldite a short length of small diameter tube to take the vacuum pipe. The take-off points need to be situated on the top of the manifold to prevent

road session with an experienced operator is a must.

A change of inlet manifold will be required to use a single sidedraught carb on a B and a choice is available from tuning component suppliers. Ideally, you should go for the longest one that fits and still allows an air filter to be used.

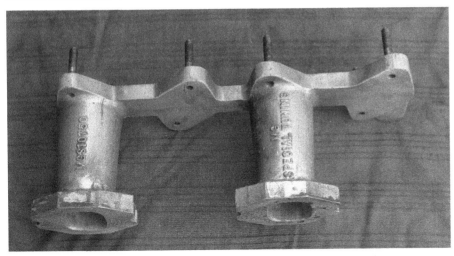

Split sidedraught carburettor manifold (straight shot at ports).

any liquid fuel present in the manifold from running into it and blocking the pipes. The take-off points need to be coupled together at a Tee, the third leg of the Tee then goes to the distributor. By linking the two points together the distributor sees a more constant vacuum signal, generated from all four cylinders, which enables it to function more smoothly. Taking a tapping from only one point means the distributor receives intermittent vacuum pulses from only two cylinders and this causes the vacuum advance plate to cycle back and forth in time with the pulses. You can actually hear the plate clacking inside the distributor when the engine is running, which is not much good for stable ignition timing or long distributor life!

If the manifold has a vacuum take-off and the engine is running a distributor without vacuum advance, this will need blocking off to prevent an air leak.

With the new manifold, a quick check is necessary before final fitment to ensure that there are no nasty steps where the carb meets the manifold and where the manifold meets the head. These are easily removed with a little grinding, as covered previously for the SU manifolds.

When fitting the carburettor to the manifold always use Thackery washers (double coil spring washers) and locknuts. Don't overtighten the carb onto the manifold: you need some play in the washers to enable them to absorb some of the engine vibrations which would otherwise make the petrol in the float chamber froth, causing the fuel metering to go out the window. The best gaskets to use between the carb and manifold are Misab plates, where the sealing ring is bonded on. The plastic double O-ring type also works well but is not as secure and can fail under extreme duress.

Single large sidedraughts on long inlet manifolds have largely superseded the old split carbs approach – where only one barrel from each of a pair of carburettors is used in order to get as straight a shot at each port as possible. The latter can still be done, but at that level of expense a mapped fuel injection system would be a better choice for road use, or even racing if the rules allowed.

FUEL INJECTION

On the subject of aftermarket fuel injection systems for the B, it is best to contact one of the approved specialist installers for all the details if you are interested. However, the siamese intake ports make fitment to and fuelling of a 5 port engine difficult. Injection systems are available as a complete package from Weber (Alpha) or Lumenition, or as just a control box (ECU) from Emerald, MBE, DTA, Motec, or Omex to name a few. Then you will need to source your own components such as injectors, sensors, high pressure fuel pump, etc, separately. For those with the desire to build their own ECU from scratch, DIY packages are available from the likes of Megasquirt. An online search will provide more details and offer up a host of alternatives.

AIR FILTERS

A considerable restriction to free breathing can occur before the air ever reaches the carburettor – at the air filter.

The standard twin-can filters do a good job of silencing the intake noise, and the paper filters do a good job of keeping the dirt out, but the small intakes on the cans seriously restrict airflow. Replacing them with good, high flowing filters will result in a 3 to 5bhp gain through improved volumetric efficiency.

The choice of filters available falls between the oiled cotton gauze type, such as from K&N, JR filters or Euro Web, or the oiled polyester foam variety from Pipercross, ITG Megaflow, Ramair or Jamex. They all function pretty much on a par with each other and I'm not bothered about slight differences in airflow capability, as they can all supply engine needs easily. Choice really comes down to which type best suits your budget or takes your fancy the most!

The cotton gauze filters (K&N solely use cotton) are strong and easily cleaned and re-oiled, so with careful looking after, they will last for practically a lifetime. Some of the other similar gauze types may use cotton or a man-made fibre

Chrome bumper standard air filters.

Rubber bumper standard air filters.

that lasts as long, but doesn't clean so easily, but they still flow air and filter very effectively, and that's all that really matters. The foam filters are slightly cheaper but need more careful cleaning and looking after, as they are more easily crushed but, again, they can last a long time if well looked after. All the varieties from reputable manufacturers are flame

retardant, so they do not present any fire hazard should the engine ever spit back through the carburettor.

In the case of the K&N filters, for the SU there is a choice between individual filters with polished stainless covers, or dual filters with polished cast alloy cover and raised MGB lettering. You can use the latter while retaining

the carburettors' original radiused entry plates, which greatly help airflow. The radius plates cannot be kept if individual carburettor filters are chosen, though, and you will have to get a pair of the excellent K&N stub stacks instead (they may already be included with the filters, but ask anyway). The stub stacks may look rather a peculiar shape but they flow at over 95 per cent efficiency and are essential for smooth passage of air into the carb.

With other brands of filters, use the original radius plate if necessary or, if not, go for the stub stacks instead. They may already have some means of smoothing air entry into the carbs included as part of the backplate, but you will lose airflow if they don't.

For sidedraught carbs, ensure there is enough room to fit at least the short ram pipes and give a minimum of 1 inch (25mm) clearance between them and the top cover.

Don't even consider running a car on the road without filters. The muck and grit sucked into the engine acts like grinding paste and can reduce engine life expectancy to less than a quarter of normal. Running open, unfiltered intakes might be considered clever on the race track, but even mega budget race teams run filters on their engines nowadays to protect what is already a hefty enough investment without unnecessarily shortening its life, or losing the race due to wear reducing the power output (yes, it's getting that critical!). Cheap pancake filters with 'seat cushion foam' elements do not keep out the dirt, and in most cases restrict airflow as much as the standard factory filters without filtering anywhere near as well! Running with a metal gauze or wire mesh over the intakes is even worse as this doesn't filter out the dirt but does reduce the area of the intake, so lessening airflow and, in some cases, affecting fuel metering and delivery characteristics of the carbs. So the recommendation is to always use air filters and get them from a reputable manufacturer.

Fitting the high flow filters means the engine does not have to pull so hard against a restriction. That restriction originally caused slightly more fuel to

Individual K&N filters.

be pulled from the jet and, as standard, the mixtures were correct for the engine's operating range. Without the restriction to pull against, the mixtures will become slightly weak when the high flow filters are fitted. To supply the correct amount of fuel to go with the improved airflow, across all engine speeds, the standard 1.5 inch SUs will need a change of needles. The HS4 SU carburettors fitted to the chrome bumper Bs – with the 'fixed' needles and hexagonal nut mixture adjustment – will need the standard FX needles changing to No. 6 needles. The later (1971 on) HIF4 carburettors – with the screwdriver mixture adjustment and 'swing tip needles'- will, in 90 per

cent of cases, need the standard AAU needles changing to ABDs. The other 10 per cent will need AAAs. The mixtures will need checking on a rolling road to verify that the fuelling is correct at all speeds, and allow adjustment should the engine be one of the odd 10 per cent that the ABDs are not quite correct for. The change of needles does not mean fuel consumption will increase; the engine's improved breathing actually means that power is produced with less throttle opening necessary than was previously used, so consumption can actually improve. That is as long as you don't enjoy driving the car harder just to hear the gorgeous intake growl from the carbs too often!

Filters with alloy MG cover and stainless steel filter for sidedraught.

FUEL SUPPLY

As already mentioned, when the engine has been modified to produce a lot more power than standard the original fuel pump will not be able to supply sufficient fuel to keep up with the engine's needs. Details of how to calculate its new supply requirement needs are given in Appendix I.

Good quality pumps from FSE (Facet), Mitsuba, Aeromotive or Holley, to name a few, are readily available, but make sure they use the correct polarity (+ve. or -ve. earth) to suit your vehicle. They must be mounted near the fuel tank in order to function correctly, as they work properly when pumping the fuel to the carbs from the tank and not when having to suck the fuel from the tank over any great distance. Put a small filter between the tank and the pump (make sure it offers no restriction to flow) to keep any debris from the tank from damaging the pump.

To go with the uprated fuel pump an adjustable pressure regulator will also be needed. Too much fuel pressure makes it hard for the carburettor float needles to do their job of governing the fuel supply, and if they are overpowered the carbs can flood. Ideally, a fuel filter should also be included to keep muck and grit from blocking the jets. Filters and regulators are available from the same manufacturers as the fuel pumps, as well as from the likes of Malpassi or Purolator, or as a combined unit such as the Malpassi Filter King. They can be mounted in any convenient, safe location under the bonnet.

When using an uprated pump to supply SU carburettors, the fuel pressure needs to be set to 1.5psi.

When supplying Weber or Dellorto sidedraught carburettors, they require a fuel pressure of between 3.5 and 4psi.

You will need a pressure gauge to correctly set the fuel pressure. Connect it to the feed pipe for the carburettors and adjust the regulator to the required value with the pump running and the engine off. Or the pressure gauge can be Teed into the fuel supply line and used with the engine running. Watch out for fuel squirting about when the gauge is disconnected.

Fuel for thought!

Finally, a few extra points are worth covering.

Always ensure the flexible fuel supply lines are petrol proof as well as correctly fitted and secure. Flames look all right only as paint jobs on custom cars.

Fuel chemistry is constantly evolving in an effort to comply with increasingly stricter modern emissions regulations. Reformulated fuels available in the US can contain up to 15% ethanol. Ethanol attacks rubber, and so can some of the other aggressive additives included in modern fuel blends. Rubber fuel hose and some older gaskets used in fuel pumps and carburettors can be susceptible to damage from prolonged exposure to modern fuel formulations.

We have experienced new fuel hoses that have swollen and gone soft over a quite short space of time after fitment when exposed to petrol. In one instance investigating a strong smell of petrol revealed the supply pipe to the carb had softened to the extent that it had split around the hose clamp.

Fuel vapour also permeates rubber hose. It may explain the slight petrol smell noticeable in the garage – that or the lawnmower.

It is a good idea to inspect all fuel hoses and fittings as part of routine maintenance. Older braided hose can hide a rubber core, so while externally it still looks good internally its function may be compromised.

Modern carburettor gaskets are designed to cope with modern fuels, and therefore deterioration over time should be minimal. Fuel hose should be to SAE standard J30R9 or the further improved J30R12. Braided hose from a reputable brand name manufacturer is nowadays lined with polytetrafluoroethylene (PTFE); a stable and durable polymer which can easily withstand aggressive fuel additives.

Some very old carburettor floats can be made from plastic, which may deteriorate over time. Brass floats and the plastic ones used in newer carb's

or provided in rebuild kits have no such problems. Likewise, no issues with brass float bowl needles, jets, dashpot needles, etc.

Always ensure the drain pipes from the SU carburettor float chamber breathers are fitted, and are routed away from the hot exhaust manifold that sits right underneath them. Don't have them positioned so that they empty onto the exhaust pipe either.

Always check that at full throttle on the accelerator pedal the throttle linkage actually does fully open the carb butterflies. Try and adjust the cable length to give full throttle with the pedal on the floor, or fit some kind of positive throttle stop so that over-enthusiastic use of the accelerator pedal doesn't overstress the linkage or snap the cable.

If you have no problems, leave the carburettors well alone. Taking them completely to bits for the sake of routine maintenance is not a good idea.

If you are experiencing problems, ensure the rest of the engine is performing well and is in a reasonable state of tune first. Faults blamed on the carburettors are frequently not down to the carburettors at all. A full diagnostic with a rolling road session may be worth considering, or follow the stages in Chapter 14.

If you are going to take the carburettors apart get hold of the relevant manual first. There are many superb specialist books widely available. Each carburettor type usually has umpteen different variations on the same theme, most of which will warrant specialist advice or techniques that the manual will cover. If in doubt make sketches as you go along to assist reassembly.

Have any necessary gaskets ready beforehand; general reconditioning sets are available from carburettor specialists.

Clean the outside of the carbs thoroughly before dismantling, as dirt and grit are the worst enemies of these precision mechanisms.

Carbs are fairly delicate instruments, so do not use excess force for fear of stripping threads or distorting components.

A fuel filter in the supply system will help to keep any muck from blocking the jets.

Never assume that any carburettor, new or used, is calibrated spot-on for your engine as purchased. This is very rarely the case. To achieve the optimum performance from the carburation you need a rolling road dyno session.

When using a non-standard carburettor a bad throttle linkage can ruin the entire installation, making setting up a nightmare. Superb linkages are available from Moss, Mangoletsi or Magard.

Always use good, high performance, high filtration air filters to minimise airflow restriction.

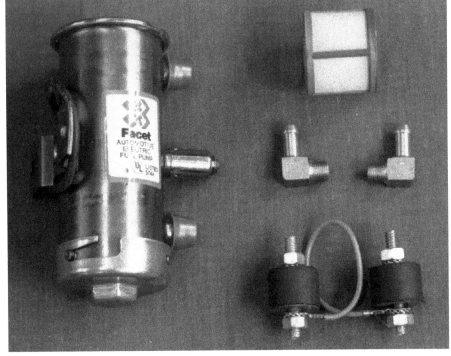

High performance fuel pump.

Chapter 9

The ignition system

The ignition system initiates combustion of the air/fuel mixture in the cylinder. However, the entire charge doesn't burn at once (the bang part of the four stroke analogy should read burn), so the engine needs the mixture to begin burning just before the piston reaches TDC. That's why ignition timing is described in terms of advance, ie 14 degrees before TDC.

The reason for the advance curve is because each rpm level has an optimum setting for performance. Assume it takes a certain amount of time for the mixture to burn and, ideally, the burn should achieve maximum cylinder pressure just after TDC on the power stroke. At low rpm there is plenty of time for the burn to happen as the piston speed is low, so little spark advance is needed (say, 14 degrees BTDC). At high rpm the piston speed is much higher and consequently there is much less time for the burn to happen,

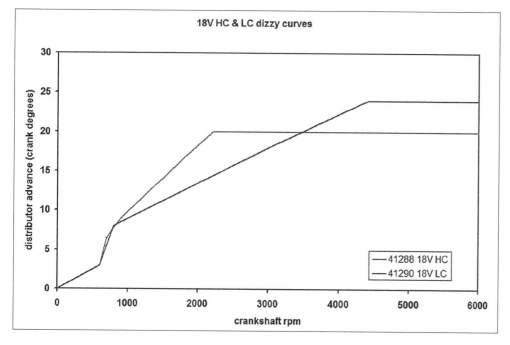

18V HC & LC dizzy curves

(legend: 41288 18V HC; 41290 18V LC)

x-axis: crankshaft rpm
y-axis: distributor advance (crank degrees)

Averaged mechanical advance curves for the 18V High (HC) and Low (LC) compression distributors, showing a faster rise and less total advance for the HC version (pale line).

Chrome bumper distributor showing 10 degrees advance plate (20 degrees at crank as distributor rotates at half engine speed).

so to achieve optimum cylinder pressure the spark must initiate the burn earlier (say, 34 degrees BTDC). The vacuum advance adds extra ignition timing at part throttle for better fuel economy and performance.

In providing the advance facility by mechanical means (inside the distributor), a slight compromise is required. The optimum performance setting cannot always be achieved for all engine rpm, though the factory is usually pretty close. Obviously, with any major engine changes these optimum settings will also change. Fortunately, MGB engines are fairly insensitive to small changes in ignition requirements and the standard distributor, in good condition, can cover most road-going applications.

DISTRIBUTORS/TIMING

In good condition the standard chrome bumper distributors are adequate for all tuning stages apart from full race. The rubber bumper distributor, however, has two problems, even for standard applications.

Firstly, the mechanical advance curve is wrong. At idle the timing is a bit too retarded, while at high rpm there is too much advance. If the distributor is retarded to achieve optimum performance above 3000rpm, throttle response and power below 3000rpm suffers. Conversely, if the timing is advanced at idle to produce a crisp throttle response and power below 3000rpm, the engine loses power above 3000rpm. The extra advance may also

Rubber bumper advance plate showing 15 degrees advance (30 degrees at crankshaft).

cause detonation which could damage the engine.

Secondly, the vacuum advance is wrong. On the chrome bumper cars the vacuum advance take-off is positioned on the carburettor, upstream of the throttle butterfly, so there is no vacuum signal at idle, the distributor advance being purely mechanical, and as the throttle opens and the vacuum take-off hole is progressively uncovered, the depression across the hole increases which, in turn, pulls more and more vacuum advance. This is good for part throttle economy. On the rubber bumper car the vacuum take-off is on the inlet manifold which is downstream of the throttle butterfly. This situation generates maximum vacuum at idle, rapidly diminishing with increasing throttle opening, which is not good for economy. On rubber bumper cars the number of degrees of vacuum advance from the distributor is far greater than on chrome bumper cars, and would cause 'pinking' if fitted to an upstream vacuum take-off position.

The contact-breaker points never seem to last very long in MGB distributors which, coupled with distributor shaft bearing wear, causes the distributor to become inefficient with age. However, as long as the mechanical and vacuum advance mechanisms are still working correctly, it is well worth fitting an aftermarket electronic ignition system (which works

Lumenition optically triggered ignition system.

Aldon Ignitor hall effect (magnetically triggered) ignition system.

well in spite of any spindle wear) to do away with the points. A wide variety of excellent systems are readily available from the likes of Lumenition, Accel,

Mallory, MSD, Piranha, Micro Dynamics or Aldon.

For those who favour originality and don't mind setting breaker points, or have

engine. Their rapid deterioration causes arcing across the points' gap, in turn causing them to fail. We now make our own, using large high quality capacitors which mount alongside the coil, external to the distributor.

When it comes to buying a new distributor, most parts and accessory specialists carry a range of 'tuned' examples. As long as the mechanical advance curve is set to give around 20 degrees maximum at around 3000rpm, it will be suitable for 99 per cent of all applications. Typically, the distributor should be fitted to give a static timing of 14 degrees at idle, giving a total of 34 degrees maximum (14 degrees at idle plus 20 degrees mechanical advance inside the distributor equals 34 degrees total timing) for both leaded and super unleaded fuel. For cheaper (lower octane) unleaded, set it to 11 degrees at idle (31 degrees max). For road use it is best to retain a vacuum advance facility to give good part throttle cruising economy, one with similar characteristics to the chrome bumper type is best.

For race use with wild camshafts and large carburettor chokes, the engine runs very poorly at low rpm. This situation can be improved by having more advance at low rpm, typically 20 degrees at idle with a mechanical advance of 10 degrees (giving 30 degrees total timing). Here, the vacuum advance can cause slight fluctuations in ignition timing, so it's best to run the distributor without it.

IGNITION COILS

The standard factory coil is of the high output variety and is adequate until it wears out, when it is best replaced with a sports coil, eg the Gold Lucas Sports or Aldon/Pertronics flamethrower coil. For post-1975 cars the Ballast resistor must be bypassed in order for these to function correctly, which usually entails

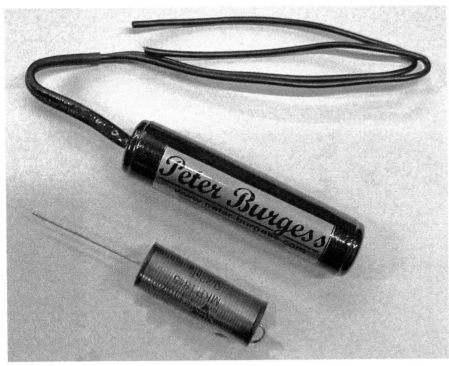

Our own replacement points condenser is much larger than the original, so requires mounting externally to the distributor ...

... somewhere alongside the coil is usually the most convenient.

to run them due to class regulations, one growing issue is the increasingly poor quality of condensers available. Many

are very badly made, suffering internal shorting and overheating as well as being unable to withstand vibration from the

Sparkplugs. Top left: EV type race plug; top right: standard plug; centre: 'V' groove lead-free plug.

running a new 12V supply directly to the coil from the ignition switch, totally bypassing the original power feed cable.

SPARKPLUGS

There are many different types and makes of sparkplug available. I always use NGK which seem slightly above average in terms of power output, length of life and resistance to fouling, but the equivalent types from other manufacturers are also satisfactory. It depends upon personal preference which you choose.

For road use where the car is likely to see mainly town and urban use or predominantly short journeys, use BP6ES, which run hotter and so keep cleaner (burn off combustion deposits) with this type of motoring. For a car that

is used mostly for longer journeys, such as open road countryside and motorway driving, use BP7ES.

For race use fit BP8ES, BP7EV or BP8EV plugs; it all depends on the engine specification and which plugs produce the best performance. The EV type of plug uses a very thin platinum centre electrode that requires only a low voltage to create a spark and functions well in an engine with high compression pressures, but it is expensive and has a short life – only four or five ten-lap races!

All the plugs usually perform best with a 35 thou (0.9mm) gap.

SPARKPLUG (HT) LEADS

The best plug leads to use are the ones with the long sparkplug boot that fits tightly and covers the entire body of the

plug. These are better at keeping out moisture and dirt, both of which can create an easier path for the spark to take – down the outside of the ceramic and not through the plug as intended. Otherwise, all that's necessary is to keep the white ceramic of the plug body clean and dirt-free, and the same for the plug leads (a little hand cleaner and a cloth works wonders on the leads), to reduce the chance of spark leakage.

ELECTRONIC SYSTEMS

There is an alternative to the standard mechanical distributor in the form of the electronic 123ignition MG distributor. This self-contained unit provides 16 pre-programmed timing curves which are selected through a small rotary switch on the underside. The various curves

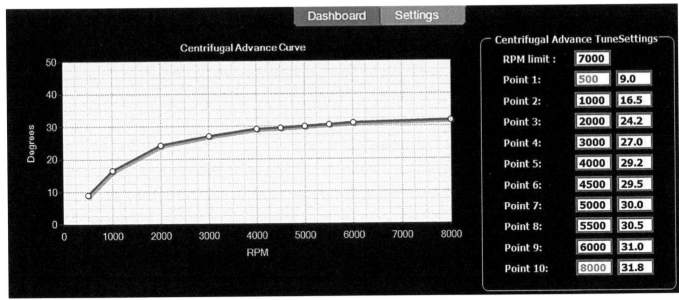

An example ignition timing curve from the 123ignition electronic distributor.

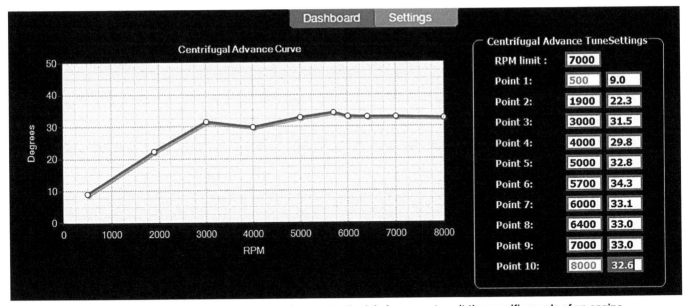

Using the 123T software, changes can be made to a standard timing curve to suit the specific needs of an engine.

can have a fast or slow rate of rise and give more or less total advance, so one best suited to a particular state of engine tune can be found. An initial setting can usually be recommended by the supplier, but this is where a rolling road session would be of benefit to try some of the alternatives.

The electronic 123Tune offers the ability to further optimise the ignition timing at discreet rpm. By connecting a laptop to the distributor the shape of the timing curve can be varied at specific rpm using the supplied software. This could mean advancing or retarding the timing at discreet points; additions or

subtractions not achievable with linear advance curve distributors.

For the ultimate in ignition systems you could go for the crankshaft triggered programmable, mapped ignitions that use one ignition coil per cylinder, which are becoming widely available. But, as for a fuel injection system, it will need tailoring

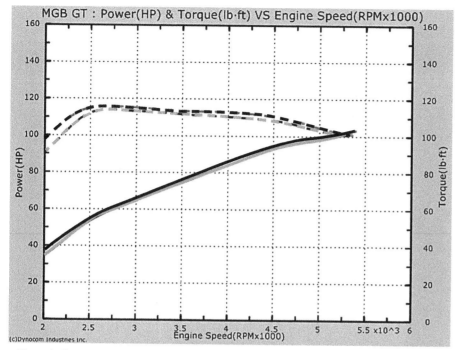

MGB GT : Power(HP) & Torque(lb·ft) VS Engine Speed(RPMx1000)

(c)Dynocom Industries Inc.

The pale lines show power (solid line) and torque (dashed line) at the wheels using the best pre-set advance curve available from an electronic 123ignition MG distributor. The dark lines demonstrate the gains offered by the capability to further adjust the ignition timing at various discreet rpm using the programmable electronic 123Tune distributor. Although maximum power output from both distributors is the same, at all other rpm power and torque is improved with the optimised ignition timing of the 123Tune.

specifically to your engine, so you're better off researching them on the internet and talking to the relevant specialists.

DETONATION AND PRE-IGNITION

While on the subject of ignition, the causes and effects of detonation and pre-ignition are now discussed as the ignition timing has the greatest effect on the problem.

Under normal circumstances the spark ignites the mixture in the combustion chamber. The burning of the mixture starts with a small kernel of flame which steadily expands and spreads throughout the chamber. As the flame advances across the combustion chamber, the unburned gases ahead of the flame are subject to increasing

temperature and pressure. The last part of the mixture to burn is called the end gas.

Detonation is the uncontrolled explosion of the end gas as it ignites spontaneously due to excessive heat and pressure in the combustion chamber, and is characterised by a 'pinking' metallic sound coming from the engine. The noise is caused by the shock wave from the explosion passing through the engine block. The energy released during severe detonation is enough to knock core plugs out of the block (especially the middle and rear plugs on the sparkplug side of the MGB block), or even punch holes clean through the top of the piston. Light detonation can slowly nibble down the side of the piston, leaving the rings exposed (which burn away due to pre-ignition).

Detonation is usually caused by having too much ignition advance or too high a compression ratio. A weak mixture doesn't help the situation but, on its own, does not usually cause detonation.

Pre-ignition is when the mixture in the combustion chamber is ignited from a source other than the sparkplug. It is caused by something glowing red hot in the combustion chamber area, such as sharp edges left after skimming the cylinder head or block, or by an overheated exhaust valve. Occasionally, sharp edges left after fitting valve seat inserts can also be the culprit. It is best, therefore, to deburr anything that feels sharp before assembly to preclude this happening. With an old, high mileage engine (or a newer engine used for short journeys with continual cold starts), pre-ignition can be caused by carbon build-up on the piston crown or combustion chamber surfaces. This is exacerbated by lead-free fuel, due to the very hard combustion products deposited in the chamber and on the piston crown. The only solution is to de-coke the cylinder head and the top of the piston, but don't remove the carbon from around the edges of the piston as this acts as a compression and oil seal on very worn engines!

Occasionally, the extra heat build-up caused by detonation can lead to the onset of pre-ignition.

The pre-ignition that most people experience is when the engine runs on after the ignition is switched off. On an MGB this can be caused by a high idle rpm, a weak mixture or retarded ignition timing, or even all three. The problem is most commonly caused by excess heat in the exhaust valve when the engine is at idle, which is part of the reason for using bronze guides on lead-free conversions in order to remove excess heat and reduce the likelihood of this happening.

Chapter 10

The exhaust system

The exhaust system can be divided into two components that have separate functions but which affect each other.

EXHAUST MANIFOLD

The first part of the system is the exhaust manifold (header). Its job is to contain the exhaust gases that leave the engine after each power stroke. These gases leave the exhaust ports as high speed, high pressure, intermittent pulses. In front of each pulse is a high pressure area, greater than atmospheric pressure, and behind each pulse there is a low pressure area, which is less than atmospheric. The low pressure helps draw out the remaining exhaust gas from the cylinder.

Also created when the exhaust valve cracks open are sudden pulses of energy in the form of pressure waves. These waves travel at the speed of sound, passing through the pulses of exhaust gas The outgoing high pressure waves are initially contained within the individual exhaust pipes. When pressure waves hit something solid they bounce, or are reflected off it. When the high pressure wave pops out of an open pipe, or suddenly passes from a small area into a larger area, a low pressure wave is reflected back up the length of the pipe. So these waves, high and low pressure, are whizzing back and forth in the pipe regardless of which way the gas is flowing.

The idea is to time the arrival back at an exhaust valve of the returning low pressure wave in order to make use of it. These waves are only generated at the exhaust valve, so the time they take to arrive back depends on the speed of the wave, and how far it has to travel along the pipe before being reflected.

By coupling together the pipes from the various cylinders at a common point (which, in effect, creates a change in area), the low pressure waves created from the individual pipes can reflect back up all the other connected pipes and broaden the speed range of the effect.

In theory, the low pressure wave can be used to increase efficiency of the engine in two ways. Firstly, if an exhaust valve opens to a low pressure area in the exhaust pipe, then the flow rate through the exhaust port is higher, which reduces the engine's pumping losses. Secondly, during the overlap period when the inlet and exhaust valves are open together (piston near TDC at the end of the exhaust stroke), the low pressure created in the cylinder and exhaust port by the rapidly departing gases will help start the fresh charge flowing sooner through the inlet valve. This increases the cylinder filling and improves engine volumetric efficiency.

Although the flow of exhaust gas is not steady, the time taken for these pressure waves to reflect back and forth is fairly constant, so the effect will occur over only a very narrow range of engine speed. On an engine equipped with a separate exhaust port per cylinder,

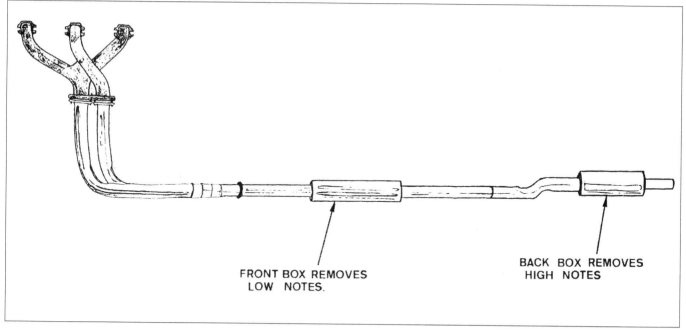

FRONT BOX REMOVES
LOW NOTES.

BACK BOX REMOVES
HIGH NOTES

Standard MGB exhaust system.

quite a lot can be done to tailor engine power delivery using these principles, by playing with the lengths of the exhaust manifold pipes. The first pipes that make up the manifold are called the primary pipes. These can join with further individual pipe lengths, called the secondaries, which finally combine at a collector, just before reaching the exhaust system itself.

Broadly speaking, there are three types of exhaust manifold layout. The first type uses very long primaries to keep each cylinder's exhaust pulses separate from the others. This pipe length is called the manifold's tuned length (approx 36 inches (914.5mm) from exhaust valve to collector for a B). The intention is to generate a low pressure signal in each pipe, which arrives back at the exhaust port just as the valve opens for the next exhaust stroke. All the individual pipes come together at a collector, no secondaries are used. This 'independent' manifold design (called a four into one manifold

on most four cylinder engines with separate exhaust ports) functions best at high rpms and is most commonly used for racing.

The second type of manifold has much shorter length primary pipes (approx 18 inches/457.2mm) which are paired together – usually cylinders one and four and cylinders two and three – into two secondary pipes (which are again approx 18 inches long/457.2mm) that finally come together at the collector. The entire manifold length is usually similar to that of the long primary 'independent' version. In theory, the exhaust pulse from number one cylinder is helped along by the low pressure area in its primary generated by the earlier exhaust cycle of number four cylinder, and vice versa. The same helpful effect occurs between cylinders two and three. This design is known as an 'interference' manifold (or four into two into one, or tri-y design), and its effect is to enhance mid-range power. This is ideal for a roadgoing engine.

The third type of manifold evolved from the motorcycle world and is intended to combine the best of both worlds. The primaries are very long, as per the independent manifold, but they then connect to a pair of secondary pipes, again similar in length to the interference manifold, before the collector leading to the exhaust system. This type of manifold design is still somewhat uncommon in the car world.

The diameter of the pipes can also influence these characteristics. Small bore pipes, as used on road manifolds, keeps the gas velocity high in the pipe and so help produce a stronger negative pulse in the system. Larger bore pipes are generally used to allow the greater volume of exhaust gas produced at high rpms to flow more easily, whilst still maintaining a similar pulse assistance effect as the smaller bore pipes.

As you have probably gathered by now, there are many combinations of design, pipe length and pipe diameters to juggle with. The only problem with all

Standard cast iron exhaust manifold.

these tuned lengths and pipe sizes is that they only have an affect over a fixed, rather narrow, rpm band. The specific tuned length varies depending on engine size and the rpm at which the effect is wanted. At another rpm range the effects may be all wrong and the engine can lose power, experience flat spots or hesitation whilst accelerating. Throwing in the effects of engine modifications, such as head swaps, cam changes, etc, adds yet more variables for inclusion in the exhaust manifold design equation. Coming up with a design to suit every case would be near impossible, so compromises have to be reached. These compromises are usually arrived at after a great deal of exhaustive testing (ouch!).

However, as the B series engine has the centre two exhaust ports siamesed in the head, most of the theory seems to go out of the window.

The standard manifold is somewhat of a mixture of the independent and

interference designs. The primary exhaust pipes of cylinders one and four combine approximately 18 inches (457.2mm) from the exhaust valve into a secondary pipe of roughly 18 inches (457.2mm) in length. Cylinders two and three exhaust into a primary type pipe around 36 inches (914.5mm) long. The secondary, from cylinders one and four, meets the long primary from cylinders two and three at a collector situated under the car before connecting to the exhaust system itself.

EXHAUST PIPES & SILENCERS

The second component is the actual exhaust system, the pipe that runs from the manifold to the rear of the car. This, again, has two main functions. Firstly, it enables the fitment of silencers to keep the noise generated by the pressure waves and high speed exhaust gas to an acceptable level. Secondly, it contains

the exhaust gas in a closed system until it can be released to atmosphere in such a position so as not to gas the car's occupants or present a hazard to others.

Similar effects to those described for the exhaust manifold also occur in the system itself, although they are usually created too far away from the engine to be of any real use. It is the positioning of the first silencer that has the most effect on the power curve of the engine in this respect.

Silencing the noise is usually achieved by either baffles or absorption. Baffle design silencers usually contain various obstructing plates with holes in and empty chambers that serve to break up and scatter the pressure waves until they lose their energy. The problem is that, in some cases, this tortuous and convoluted path can hinder the flow of the gas and create flow restriction called backpressure. This type of silencer is more widely favoured in the USA, where manufacturers have largely eliminated this problem.

The absorption silencer allows the exhaust gas to flow freely through it, along a straight, perforated tube. This allows the pressure waves to expand through the perforations and become slowed and diffused through the densely packed, heat resistant wadding that surrounds the tube.

The standard car already has an efficient cast iron manifold which comprises one half of the long primary for cylinders two and three, and the complete primaries for numbers one and four cylinders. Cast iron is used for its excellent acoustic absorption and noise damping ability (some tubular manifolds 'ring' under certain conditions) and for its superior heat transfer characteristics (very hot exhaust ports lose power through heat transfer to inlet ports alongside). Cast iron is also well suited to handling the very hot, chemically

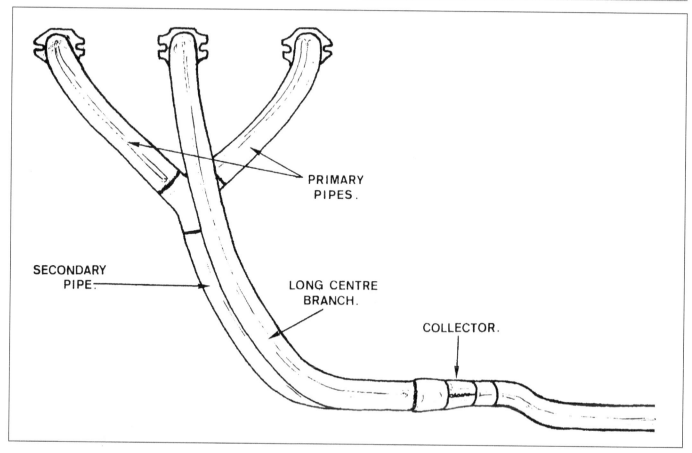

PRIMARY PIPES.

SECONDARY PIPE.

LONG CENTRE BRANCH.

COLLECTOR.

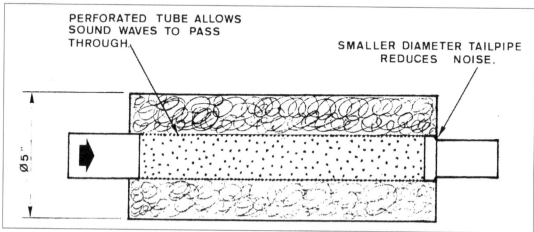

PERFORATED TUBE ALLOWS SOUND WAVES TO PASS THROUGH.

SMALLER DIAMETER TAILPIPE REDUCES NOISE.

Ø 5"

MGB LCB manifold.

Left: Standard absorbtion type silencer, which uses a perforated tube surrounded by packing material (usually glassfibre wool) to absorb soundwaves (noise).

An instant mid-range power gain of 2 to 3bhp at the wheels can be achieved by removing the first silencer box from the standard system and replacing it with straight pipe. The noise level does increase, but not to an annoying extent as the first box serves to eliminate the low frequency noises from

very unpleasant mix belched out of the engine on every exhaust stroke. Casting is also cheaper in mass-production than actually fabricating a complicated manifold shape. Having said that, the second half of the primary pipe of cylinders two and three and the entire secondary pipe section for cylinders one and four is a fabricated piece made in mild steel.

the exhaust. The result is a nice, deep, mellow-sounding exhaust note. Removing the rear silencer instead is not to be recommended as its job is to filter out all the irritating higher frequencies, the really nasty crackly, raspy notes.

The Peco exhaust manifold and system we originally recommended is no longer available. The best exhaust to date in terms of quality, effectiveness and reliability we have tested is manufactured by Maniflow. It is a slightly larger bore (pipe diameter) than standard. There is a choice of system, with a single, larger-than-standard

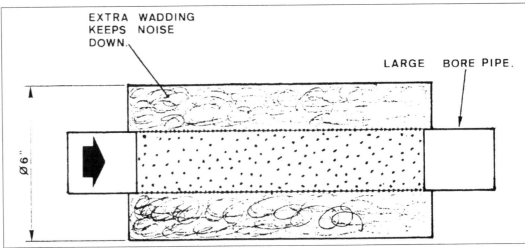

EXTRA WADDING KEEPS NOISE DOWN.

LARGE BORE PIPE.

Ø 6"

Straight through silencer.

silencer box at the rear of the car, or dual silencers. Fitting either one typically boosts mid-range power by 5bhp at the wheels. The silencing ability is excellent, too; producing a deep, powerful-sounding exhaust note, the dual system being quieter and with a more mellow tone.

Maniflow offer several different exhaust manifold designs for the B, with differences in header pipe shape for

The following photographs show examples of Maniflow exhaust manifolds and systems:

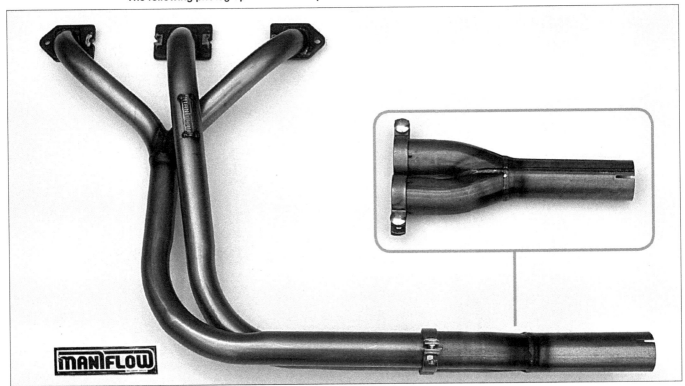

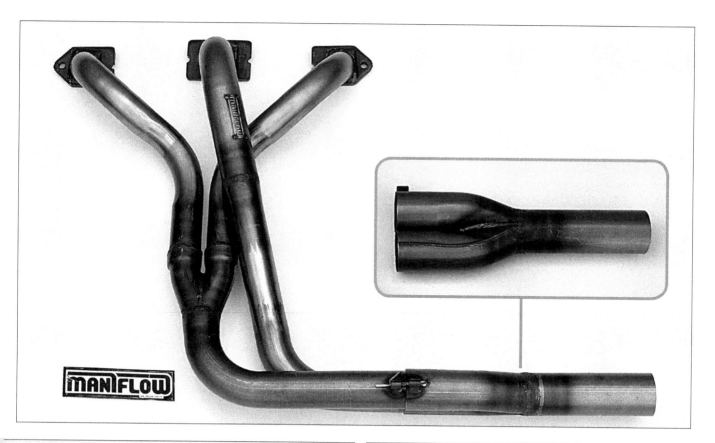

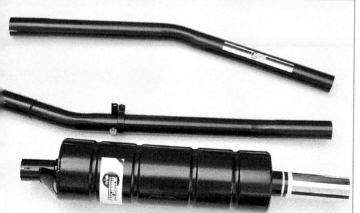

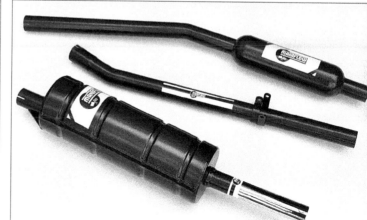

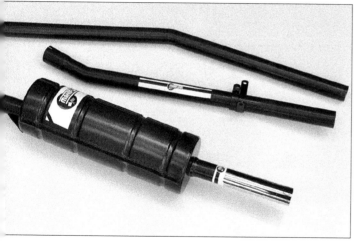

SU or Weber sidedraught carburettor applications. It is advisable to contact the company directly should you wish to purchase one, as it is happy to discuss your requirements and recommend a suitable manifold and system for your specific application.

On the standard race class MGBs, where the exhaust manifold has to remain standard, it may be necessary to make or purchase an adapter to step from the standard exhaust collector size to the Maniflow system. It is a point for consideration if you are planning on just using the exhaust system with the standard manifold and downpipes, as it is a necessary task if you want the performance gain from the system.

For chrome bumper models Moss offer a polished stainless steel exhaust system under the name Tourist Trophy. The system uses a single large silencer box and suitable downpipes to connect the standard cast iron manifold, or there is the option of a matching three-into-one tubular manifold – we have had good results from these on the dyno, and the sound is not obtrusive.

The wadding that makes up the sound absorption layer in the silencer box also plays another important, power-related role, as well as quietening the exhaust. If the back box becomes hollow – due to deterioration of the wadding

and it being blown out the tailpipe – the car looses up to 12bhp at the wheels! Oddly enough, this only shows above 4500rpm when the engine goes flat. With the car on the rollers the engine refuses to respond to fuel or ignition timing changes. The exhaust note will also be noticeably louder, though you may not have noticed when driving as noise increase would have been gradual as the silencing effect of the wadding decreased. A quick method of checking is to tap the box with a metal rod/bar or a large spanner: if it's empty it will obviously sound hollow.

I mention this problem as it is a very common occurrence on MGBs, especially with some stainless steel silencers that are fitted. Luckily, these are guaranteed for life and should be replaced in order to regain the car's lost sparkle.

Moving onto the subject of race MGBs, for full race use I would recommend using a long 3 into 1 exhaust manifold (independent), coupled to the Maniflow system. Chat to Maniflow regarding your specification and requirements.

As a bit of an aside while on the subject of race Bs, we ran a championship-winning car – in the standard race class – for a whole season using exhaust insulating wrap. The

thinking was that the engine bay would run cooler, as the exhaust heat was now contained within the pipes, and so would allow the engine to make more power from inducting the cooler, denser air. After fitting the wrap no difference in power was recorded with the car on the rolling road, but we left the wrap on as we thought that under actual race track conditions, with the very high underbonnet temperatures that racing produces, there would be more power available. The actual theory of reducing the intake air temperature is correct: every 3 degrees C reduction gives about 1 per cent more power. Unfortunately, with hindsight, I can see that we went about it the wrong way.

After the end of the season when it came time to fit the latest specification engine ready for the next season, the cast iron exhaust manifold would not fit back on the head. It had warped by around 0.25 inch (6.4mm) due to the extremely high temperatures it had experienced. I had created my own problem. On the one hand, the underbonnet temperatures were lower, so perhaps giving a little more power, whilst, on the other hand, the engine was actually losing power by overheating the exhaust internals as a result of trapping all that extra heat with the wrap. This made the cylinder head hotter, which

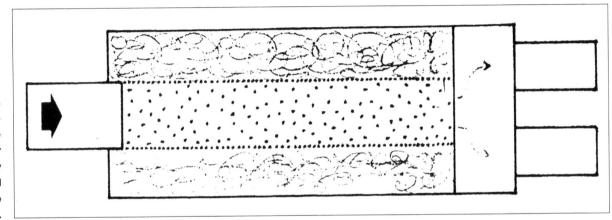

Stainless steel twin-tailpipe-type silencer can be very noisy and may lose power.

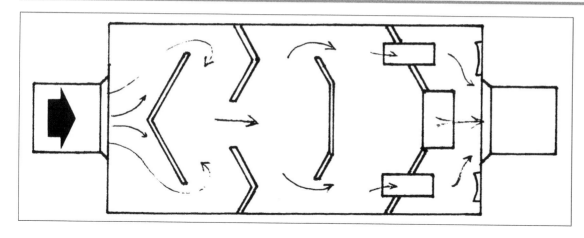

Reflective silencer splits exhaust gas flow path and uses plates to reflect sound back at itself as a noise elimination method.

heated the inlet ports and then the fresh mixture. So, scrap one twisted exhaust manifold, time for a rethink. The simple solution was to fit a thick aluminium plate as a heat shield in order to physically separate the inlet and exhaust manifolds. This shield then allowed each component to function as it was originally intended.

Further testing on another car (not a B) fitted with a Ford Crossflow engine – where the inlet is on the opposite side of the head to the exhaust so heat doesn't effect it so badly – the car showed a 3bhp at the wheels gain once the wrap was removed. Make of it what you will, but it appears not a good idea to discard the standard heat shield from the car: it was put there for a purpose. Maybe alternatives such as lagging the pipes

had been tried and rejected by the original designers. Who knows?

If you fancy making your own exhaust manifold for a roadgoing car, there is one further variation of manifold design – as fitted to MGAs – that gives more mid-range power and yet loses nothing at high rpm. With this manifold all three pipes join up approximately 18 inches (457.2mm) away from the exhaust valves and then run as a single pipe to halfway down the engine block. This adds 2 to 3bhp more at the wheels to mid-range power. So, reproducing this layout with a homemade tubular manifold joined to the single box Maniflow system could increase mid-range power by some 7 or 8bhp at the wheels without significantly increasing noise levels.

If you're really keen to have a go at making your own manifold and system, then I would recommend investing in one of the specialist books available about exhaust system design. It is really quite a complicated subject which goes far beyond what is covered here. Exhaust components, pipes, bends, silencers, etc are readily available from the likes of Jetex should you wish to make your own manifold and system; your best bet would be to use the original as a template and go from there. Even then it will take an awful lot of trial and error, as well as a lot of engine dyno time, to come up with a successful design. But then, who knows?, you may unleash the secret of gaining real power from the MGB exhaust system!

Chapter 11
Engine lubrication & cooling

ENGINE LUBRICATION

Oil is the lifeblood of any engine and shortcomings in the lubrication system will have catastrophic results. Oil protects the engine against wear, reduces friction, removes waste products, contributes to cooling and helps the piston rings seal against the bores.

In the case of the B series engine, my recommendation is to use a good quality 20W-50 oil (price is usually a reasonable indicator of quality). Oils of a wider viscosity range and with other attributes – such as synthetics – are fine if the other attributes are being used; if not, why pay the extra?

Of more importance is the frequency of oil changes. With so many MGBs used only during the summer months, and consequently clocking up relatively few miles, many owners fall into the trap of changing the car's oil only when the appropriate mileage limit is reached. Remember that the official recommendation for the MGB Is an oil change at 3000 miles or *every three months*, accompanied by a filter change at every 6000 miles or *six months*. Oil performance has improved since the time these recommendations were drafted, so doubling the oil change interval to 6000 miles or so will not harm. Although low mileage use, especially in short journeys, still gives the oil a hard time, and, with a new filter costing so little, there's everything to be gained by changing it every time the oil is changed.

For winter storage, if the car is to remain unused, renew the oil and filter. Disconnect the ignition and crank the engine over on the starter until good oil pressure is achieved (temporarily removing the sparkplugs will speed things up). Reconnect the ignition but do not start and run the engine after this: the intention is only to coat interior surfaces of the engine with clean, fresh oil which will protect it whilst it stands over winter.

The economics of using high-cost fully synthetic oils has to be evaluated. There is no doubt that it retains its 'performance envelope' longer than ordinary mineral oils, but it can cost significantly more. Also, using synthetic oil in older (design and/or mileage) engines can lead to increased oil consumption from leakage past seals and gaskets. However, fully and semi synthetic oils of the correct viscosity (15w-50 or 20w-50) specifically formulated for use in older 'classic' engines are readily available and recommended if you don't mind spending the extra. If you are not using oil formulated for classic engines, consider using an additional anti-wear additive containing zinc (as ZDDP).

Other considerations to do with the lubrication system can be just as important. Firstly, many individuals will rebuild and tune their engines once the original has worn out or gone bang. They will go to great expense having the engine rebuilt to a very precise specification, only to reconnect the original oil cooler – any debris in there will write off the engine within the first few miles! Whilst flushing through the cooler and pipes will *probably* eliminate potential problems, why risk it? These components should be renewed, especially if the original engine met its end in a destructive way.

Oil coolers

The whole subject of oil coolers and whether they are required has been the topic of considerable debate for many years. When the MGB was introduced in 1962, the oil cooler was standard for certain overseas markets and an option for the UK market. With the introduction of the five main bearing crank in October 1964 (18GB on), it was standardised for all cars. In contrast, the good old 'Land Crab' (the Austin/Morris/Wolseley 1800), which used a similar version of the five bearing engine, did not have an oil cooler. Even the S versions – which had a slightly higher tune specification to the MGB – didn't have one. This continued until the 1970s when neither version of the 1800cc Marina had an oil cooler.

The move from standard to more powerful specification engines reinforced the argument for the use of an oil cooler. However, the problems for MGBs are twofold: firstly, overcooled oil can increase wear and, secondly, an MG engine bay minus the familar cooler is regarded as incorrect, and especially so in Concours circles.

The compromise solution suitable to both problems is to fit an oil thermostat. This is a temperature sensitive valve which will ensure oil does not pass through the cooler until it has reached the correct temperature. From then on the valve opens and closes with temperature variations in exactly the same was as the water thermostat does.

Oil 'stats are readily available from a number of sources in various size bores for a variety of applications.

Baffling the sump (vertical barriers to limit oil movement) only starts to be an advantage in the arduous conditions of competition, where continuous hard cornering can uncover the oil pick-up, which leads to momentary oil pressure loss. Roadgoing cars do not achieve the same cornering forces so do not warrant this work. In the same way a baffle plate (horizontal, covering the sump well), is only of advantage on the track. Old published material on MGB tuning will also cover items such as oil pump modification raising the oil pressure. Over the years, most of these modifications were built into standard Bs, most notably from the 18V series engine (circa 1971) on, so there is no benefit to be gained by pursuing them. When the engine is fully built in accordance with guidance given within these pages, the oil system will operate efficiently long-term.

THE COOLING SYSTEM

The duty demanded of the cooling system is very simple: to maintain the engine operating temperature in a predetermined band, which will vary according to use. For this to be achieved the cooling system must have the capacity to dissipate the heat generated by the engine.

As engine power increases, the amount of heat produced rises in proportion. For almost all roadgoing conversions the standard radiator in 'as new' condition has the necessary capacity in all but the hottest climates. For very powerful conversions, hot climate conditions and competition use, consideration has to be given to increasing the heat transfer ability of the radiator. It is worth noting here that the forward-mounted, post-1976 radiators

provide better performance than earlier examples, but fitting these to the earlier cars is far from straightforward.

A problem is that the majority of radiators will have more than a few years' service under their belts. The small fin-covered tubes which allow water to flow through the radiator become clogged and blocked by silt and corrosion, impeding the water's passage and compromising radiator efficiency. Backflushing the radiator with a hose can wash out some of the loose muck, but it won't tackle corrosion or blockages, so improvement is marginal. Ideally, the radiator should be reconditioned or rebuilt, with an uprated core, if necessary. It will then be easily capable of handling the cooling of any B series engine. Many specialist radiator reconditioners can also provide custom-made oil coolers and heater cores to suit any application, which is worth bearing in mind.

The oil cooler in chrome bumper cars is mounted in front of the water radiator, restricting airflow through the latter. Rubber bumper cars benefit from having the oil cooler mounted in an underslung position, which completely eliminates this problem, although it is at the cost of slightly increased vulnerability to damage from debris; this danger can be reduced considerably, however, by fitting a steel mesh.

Chrome bumper cars can be converted to the rubber bumper car arrangement using the rubber bumper oil cooler panel and reinforcement. The reinforcement will need slight modification to account for the chrome bumper mountings, while the change of location necessitates different length oil hoses.

Standard water pumps are fine. In the case of competition engines, where the rev range has been significantly increased, a larger diameter water pump pulley will help reduce the possibility of pump cavitation. Too high an operating speed causes the impeller to 'thrash' the water instead of pushing it around efficiently. This froths the water and air bubbles, combined with reduced flow, results in overheating. Fitting a larger diameter pulley slows the pump speed, enabling it to operate efficiently, despite increased engine rpm. The exact details of this should be discussed with your engine builder who will be in the best position to provide guidance.

A thermostat should be used on *all* road engines. The ideal 'stat is 88 degrees Celsius (190F). In a few competition situations the thermostat can be discarded, in which case it must be replaced with the special blanking sleeve available from Moss. For sprint, hillclimb and some race competitors, it is best to retain the thermostat to allow correct engine operating temperatures whilst driving the car to and from the venue. Running without a thermostat and without the sleeve can actually result in a *reduced* water flow rate, so beware!

Pure water is the best medium for transferring heat. The problem, however, with running pure water is corrosion (the hotter the water the greater the corrosion) and how to cope with temperatures of below freezing. To inhibit corrosion and freezing use a quality glycol-based anti-freeze at *all* times. The actual mix will depend on operating environment and conditions but, as a guide, a minimum 25 per cent mix should be used. Remember that anti-freeze is less effective than water at conducting heat away so, ideally, the proportion of anti-freeze to water should be kept as low as possible and should not exceed 50 per cent.

Mention should also be made of the increasing number of coolant additives appearing on the market, all claiming various advantages. In our testing we found these waterless coolants increase the operating temperature by around 10 degrees centigrade, and only tend to reject heat 4/5ths as efficiently as water. We recorded a series of instances of cylinder head and engine failure,

specifically valves seizing in guides and piston seizure, which we could only attribute to an excess build up of heat. In the cases with which we have been involved, changing back to water prevented a repeat of the problem. We feel the 'window' for safe running conditions is narrowed with the waterless coolant. Excess antifreeze also has the same effects.

Finally, the cooling fan. Rubber bumper cars are fitted with an electric fan. Modified cars can benefit from the fitment of a second fan, such as used on the early V8s. Remember, though, that later V8s, which had enormous torque and hot running conditions, reverted to just the one fan.

On chrome bumper cars the fitting of an electric cooling fan as a replacement for the standard engine-driven unit is a recommended modification, be it one of the aftermarket kits or a homemade affair. The power gains are not significant but elimination of noise (the engine fan is the source of the annoying hum at speeds over 3000rpm) is!

Chapter 12
Transmission

CLUTCH

For road use it's best to retain the standard clutch. The uprated versions intended for competition are heavy to actuate and very positive in action – either in or out – making everyday driving a chore. The standard release bearing wears out rapidly, due to increased workload, and slipping the clutch results in rapid heat build-up – sufficient to 'blue' the cover assembly and damage working surfaces – as the plate drags. However, if a competition clutch is used, a stainless steel braided clutch hose is a good idea in order to handle the increased actuation pressure.

GEARBOX

The standard gearbox is fine for all stages of road and mild competition use. The 'gap' (rev drop) between first and second may be considered rather large, but the standard ratios were intended to encompass a wide variety of motoring styles and conditions, and this they do

well. If, however, you feel the scope is not best suited to your particular motoring requirements, then it may be worth considering a set of close ratio helical gears (helical gears are used in all production gearboxes because of their smoothness and quietness of operation). A change of final drive (lower) may also be called for.

An alternative that is growing in popularity is to fit a five-speed conversion, which typically uses a Ford Type 9 gearbox. The intermediate gear ratios are better spaced than a standard box, while fifth gear is the equivalent of overdrive fourth, so you get the benefits of long legged cruising while not being too tall for the engine to cope with. The high quality kit as offered by Hi-Gear Engineering provides all the necessary parts to make the swap, and does not require modification of the transmission tunnel. Alternative ratios are available to suit the vehicle and purpose, together with close ratio and competition

gearsets. Gearboxes from Japanese cars are also used by conversion specialists outside the UK, where the Ford box is not so readily available.

When an engine is modified to give more and more power, the range of rpm where the significant power output is found gets smaller and smaller, which tends to highlight the gaps between gears. In order to keep the engine in its power band, or 'on the cam,' closer and closer gear ratios are required. The rev drop between gearchanges can thus be minimised (the calculations for this are in the Appendix).

All production gearboxes are a compromise, but a gearbox meant for competition use is usually tailor-made for ultimate performance. Forget quiet, vibration-free motoring, these boxes are intended to transmit power as efficiently as possible. Straight cut gears are stronger and better able to take the abuse (occasional crunched gearchange) of competition, but they are extremely noisy in operation. These gear ratios are also intended to be used along with a change to a lower differential (final drive) ratio.

Close ratio gearsets – straight cut and helical – for competition or road use are available from Moss. Modification to the gearbox may be necessary to allow fitment.

For competition, relocating the overdrive switch to the gearknob makes operation easier in the heat of battle. An uprated overdrive unit suitable for competition use, or to give peace of mind, should you feel the original is unable to cope with increased engine performance, is available from Moss.

FINAL DRIVE

The rear axle ratios have ultimate control of the range of speeds over which the gears operate. The standard ratio for the B is 3.909:1 (3.9 revolutions of the propshaft to one at the wheels), in either the banjo (one-piece axle casing with the diff in the 'nose') or the Salisbury (cast iron centre, pressed in axle tubes with access to the diff through the cover at rear) rear ends. This ratio was chosen by the factory to suit normal driving conditions considering the power of the engine and weight of the car, as well as to minimise engine revs at speed and give good fuel economy.

Alternative ratios are available, though components from the two axles are not interchangeable. A lower ratio

(numerically higher, ie 4.55:1) effectively lowers all the gears and reduces top speed attainable in them. This will make the car quicker in acceleration but will necessitate more gearchanges. Long distance or motorway work will become increasingly 'thrashy' as the engine will have to be revved harder to achieve desired speeds, and fuel consumption will increase as a consequence. Obviously, these ratios are intended more for competition work or use in conjunction with close ratio gearboxes.

A higher ratio (numerically lower, ie 3.07:1) increases attainability of high speeds in each gear, giving greater top speed and improving fuel economy, but does reduce acceleration. If overdone, it can mean the engine struggles through insufficient torque to pull the car along at lower operating rpms.

For racing the ideal choice will allow the engine to reach peak power in top gear (or OD top) at the end of the fastest straight on the track. This becomes a little more difficult for sprint and hillclimb events as not as many gears are used during a run, but, again, maximum power in a particular gear at the fastest part of the track would be best, in such a way that the engine stays 'on the cam' for other gearchanges made. Don't forget about travelling to and from the venue, though!

Unless you are involved in competition, the standard final drive ratio is best.

Limited slip differentials are available for both axle types. Should one wheel lose traction, a limited slip differential ensures that the spinning wheel transfers a set amount of its power to the other. It can make a powerful machine more controllable when cornering and can be of some benefit in bad weather, where traction may be limited.

SPEEDOMETERS

Any change of axle ratio or tyre size will affect calibration of the speedometer. Most standard B speedo readings err on the side of optimism by around 10 per cent in any case.

Recalibration will be necessary to compensate for any major changes and, by supplying the relevant information such as tyre size, axle ratio, etc, the speedometer can be tailored to suit. The gauge can be overhauled at the same time – a wandering needle is a sign of wear (the rotating magnets which move the needle wear in service and can disintegrate) that will require rectification. A check that the nylon drive gear in the gearbox is in good condition, and that the speedo drive cable is in good condition and is kept well greased, would not go amiss.

Next time you take your car for a rolling road tune, enquire if the dyno speed readout is accurate and ask for the speedo readings to be checked against it.

Chapter 13
Suspension, brakes & tyres

'CHASSIS'

Before embarking on an upgrade, the vehicle should be mechanically and structurally sound. Remember that the suspension is connected to the car's structure, and any flexing will alter the platform the suspension works from and so alter the handling. Uprating will put greater strains and loads on the chassis and will rapidly highlight and expose any structural deficiencies.

If the car is undergoing full restoration, or if a new shell is to be used, it can be worthwhile checking for correct alignment, either by using specialist equipment found in many larger garages and body shops, or by a simple chassis drop test using a plumb bob, chalk, string and a flat surface (details for this can be found in the official workshop manuals). Don't be fooled into believing that because the car has never suffered from accident damage it will be straight: distortion can also occur through years of general use,

corrosion weakening or high mileage. Seam welding the panel joins will stiffen body structure enormously, giving a rigid base for the suspension to work from. Further structural modifications could be carried out at the same time; for example, remounting of the oil cooler to the underslung rubber bumper position or adoption of additional chassis bracing that extends from the central crossmember to rear spring positions on the MG RV8, which also form the front anchor points for anti-tramp bars which could also be utilised.

Rear anti-tramp bar conversion.

SUSPENSION

Suspension, ride and handling requirements vary enormously between individual drivers. There is a lot involved in setting up a car and it's not easy, especially when modifying a road car. To achieve the correct set-up it's necessary to ensure that all the components involved with the suspension work correctly and in harmony. The car's cornering, stopping and acceleration performance is dictated by the tyres, with a contact area for all four of approximately the size of an A4 page. The aim of the suspension is to keep the tyres in meaningful contact with the road surface.

Many owners are unaware of the true handling capabilities of the standard MGB, so the first priority must be to get the suspension back to as-new condition. This means replacing everything – front and rear springs, dampers (shock absorbers), suspension rubbers, kingpins, etc, all at once (preferably). Make sure you choose a reputable source for replacements where advice can be relied upon, as some components can be of sub-standard quality. The alignment, toe, etc, should also be checked and set correctly. The transformation in character, once all this work has been done, is surprising. Now, if you are still not satisfied with the car, you can begin to change and uprate components with this baseline setting to work from.

Handling during cornering can be categorised as neutral, understeering (where the car wants to go straight on) and oversteering (the rear of the car wants to come around and overtake the front). As new, the B handled extremely well, being very neutral when cornering and displaying slight oversteer when driven hard. These characteristics can be altered, if desired, by changing or uprating the suspension components. As a rule of thumb, stiffening the front will increase understeer/decrease oversteer, whilst stiffening the rear will increase oversteer/decrease understeer. For road use, a neutral to slight understeer condition is safest because it's a natural tendency to turn the steering wheel further into the corner.

Tyre pressures can also affect these characteristics (see tyre section) but should always be returned to normal settings before evaluating any suspension changes.

If you are creating your own specification rather than buying a full kit, any changes should be gradual (one thing at a time) so the effect of each can be fully assessed. Drawing on the experience of others can be useful. Carry out any handling tests on a suitable (large) private area, and not the public highway.

A good first improvement is to fit uprated suspension bushes (the standard front wishbone bushes are prone to collapse), removing some of the standard compliance and deflection.

Firmer bushes from a BGT V8 were a common upgrade.

An excellent alternative is to fit polyurethane bushes, which can be available in different grades of firmness, unlike the standard rubber versions. They are designed to remove unwanted deflection, and not squish and flex like rubber. Nor do they deteriorate the same, being a polymer, and offer increased longevity by also being better able to withstand the rigours of road use and associated contaminants.

There is another bushing option on the market in the form of Nylatron – a hard plastic. Once properly fitted they are very effective at improving location of suspension components as they have no compliance. The downsides are the need for fairly frequent maintenance and a marked increase in NVH – noise, vibration and harshness! The Nylatron bushes are best suited for race only, class rules permitting, where comfort and ride quality are not important factors.

Anti-roll bars act to counter body roll, their stiffness being related to their thickness (diameter), provided the material is the same. They interlink the suspension and when the body rolls during cornering they twist (a torsion bar), so helping resist the roll motion. They do not affect the spring rate (suspension stiffness) when the vehicle is travelling in a straight line. For them to work effectively the body must roll, or at least not be prevented from doing so by springs which are too stiff. Too stiff a bar, however, will cause severe understeer and can lead to front wheel lifting under very hard cornering.

The majority of MGBs are fitted with front bars of 0.56in (14.3mm) diameter for the roadster and 0.62in (15.8mm) for the GT. Alternatives of up to 1 inch (25mm) diameter are available. Only the late rubber bumper cars have a rear bar. This was necessary in order to counter some

of the awful handling characteristics introduced by the 1.50 inch (37mm) body height increase imposed by US legislation. Fitting rear anti-roll bars to chrome bumper cars is unnecessary as body roll is not a problem.

Changing spring rates is another common way of altering handling. Here, the first consideration must be the car's intended use – carrying passengers, etc. If this involves just road use then it's best to retain original ride height, keeping spring rates close to standard to retain a degree of compliance for both comfort and rough surface handling. Stiffening the spring rates will reduce body roll, although this is not a major factor in how well a car corners. Springing that is too stiff may give the car F1-type handling, but it will be a nightmare over bad road surfaces. A 25 per cent increase is a reasonable first upgrade. Stiffening the front springs will increase understeer/decrease oversteer. Stiffening the rear springs will increase oversteer/decrease understeer.

Changing springs can often be accompanied by lowering the car. Chrome bumper cars ride quite low already and the middle box of the exhaust system can be caught on obstructions or by traffic calming measures. Lowering the car further is not advised, except for smooth surface use competition cars.

Reduction of rubber bumper car suspension height to chrome bumper car level does reinstate most of the lost handling ability, though significant structural body changes were incorporated by Abingdon in order to raise the ride height of rubber bumper cars and the suspension function and geometry will be seriously affected by a springs-only change.

Lowering the car will alter suspension geometry, introducing more negative camber at the front. Negative camber is when the top of the tyre leans inward and positive when it leans outward, viewed with the car static and on a level surface. For optimum handling the car's outside tyres (loaded) should stand vertically when cornering (zero camber) so that all the tyre tread is in contact with the ground. Body roll, however, can cause a camber change during cornering. In some cases the change results in positive camber on the outside tyre during cornering, unless the suspension is designed in such a way to counter this, so producing less cornering force and resulting in understeer. The only way to overcome this camber change is to create more negative camber to begin with. This should then create the ideal zero camber condition in use.

Negative camber wishbones are available for the MGB and these give good improved cornering ability without having to lower the car or creating any of the bad effects of negative camber excess, such as tyre wear, high speed wander and reduced braking efficiency (in a straight line).

The original lever arm dampers have suffered unwarranted criticism over the years and fitting *good quality* replacements will often provide a very noticeable improvement. One of the most common criticisms of lever arm dampers is their relatively short life; however, this can be attributed to poor quality reconditioned items. Uprated versions are available which have alternative valving arrangements and these should be used in conjunction with any spring changes.

As good as the original lever arm dampers are, one option is to replace them with telescopic dampers and there are a number of rear damper conversion suppliers with proven track records. For those using adjustable dampers a common misconception is that they must be set very hard. This often

reduces handling capabilities as the suspension is then so solid. Experiment to find the best for your car by starting at the softest setting and working up. Don't be surprised to find that the best setting is very near the softest end of the scale.

Front damper conversions, and complete front suspension conversions, are more recent options. Replacement of the original lever arm dampers with a more efficient telescopic damper should result in considerable improvement, yet the offset mounting is not ideal (the damper mounting also operates in single shear; not desirable from the point of view of strength). The much more comprehensive front conversion offers the benefits of telescopic dampers mounted in the much better position of a double wishbone set-up.

Remember to always have the tracking (toe) checked after doing any work on the front suspension.

Adjustable telescopic rear dampers.

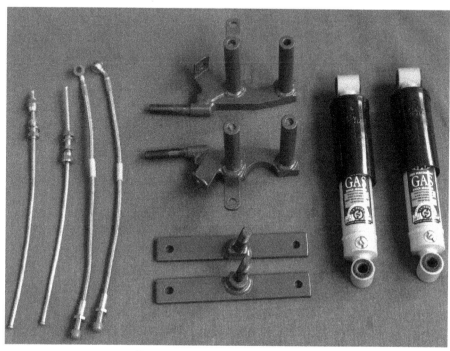

Front telescopic damper conversion.

Front 'coil over' suspension conversion.

BRAKES

The MGB is equipped with front discs of 10.75in (273mm) diameter and 0.35in (8.8mm) thickness) and rear drums of 10in (254mm) diameter. The only significant change made to the braking system over the years was the introduction of dual circuit hydraulics. All standard systems, when in first class condition, provide consistent and very effective braking.

Unfortunately, brakes are one of those oft overlooked areas of a car. They should be checked regularly to ensure there is no leakage, seizure or damage to the seals, and that they are correctly adjusted. So the first order of business is to ensure the standard brakes are working perfectly.

Everything should be moving freely and there should be no disc run-out (0.03in./0.76mm max) or wobble from the wheel bearings. Any movement of the disc means the pads are pushed further away, pedal travel increases to compensate and a judder can be felt. Discs can be skimmed to correct any problems, but if they have worn too thin (0.299 inches/7.6mm minimum thickness) replacements are necessary.

Another overlooked factor is the brake fluid. The standard glycol-based type is hygroscopic: it absorbs moisture from the atmosphere. This contaminates the fluid, reducing boiling point and seriously compromising effectiveness (the brake pedal going to the floor during racing is due to the vapour from the boiling fluid – or water – being compressible). The fluid should be renewed every year with the entire system being bled to flush the old fluid out. A fluid with a higher temperature capability, DOT 5, can be used as a replacement if necessary.

An alternative is to use silicon-based brake fluid. This is non-moisture absorbent (it should still be changed regularly, though), doesn't damage paintwork and has a higher boiling point than conventional fluid, though under high temperatures it can be compressible, giving the brake pedal a slightly spongy feel. To be totally sure of its purity the entire brake system will need flushing through with alcohol to remove all traces of the old fluid and all the rubber seals must be renewed. It's not a cheap exercise but it is worth consideration.

Rubber brake hoses could be replaced by stainless steel braided hoses, available from Goodridge, Aeroquip or Earl's. These have a longer life, are more damage resistant and do not balloon or bulge under pressure, as standard ones can due to old age. They also give slightly increased pedal firmness and marginally reduced pedal travel.

For severe road or mild competition use, pads and shoes can be changed for ones with higher temperature operating ranges. A word of caution: ensure material choice for road use is only 'one step up' from standard, otherwise reduced braking efficiency may result. Cold and/or wet weather or long spells of motorway work (little brake usage) may mean uprated pads are below optimum operating temperature, so initial stopping power will be compromised.

Generally speaking, the material of a shoe or pad doesn't function at its best until warm. That level of performance is retained as the temperature increases, before falling off (brake fade) as it overheats. Pads for road use give more bite at lower temperatures, while 'harder' competition types can take more heat before they fade. Uprated materials also exhibit a slightly lower co-efficient of friction, so need more pedal effort to achieve the equivalent braking power of normal materials.

The Mintex 171 pads and shoes are no longer available, and DS11s are too hard for road use. One alternative for sprint/hillclimb use, which is not too hard for fast road use, is Mintex MLB40M1144 asbestos-free pads. Similar pads are available from EBC, or Ferodo.

Pads may require specific bedding-in to function correctly, so always heed the manufacturer's instructions.

If the brakes are still found to be lacking after using different material, then a component upgrade must be considered, unless yours is a competition car and the rules prevent

Standard brake disc (left) and thicker V8 disc.

this. One upgrade is to fit the thicker discs (0.5in./12.5mm) from the MGB GT V8 to aid heat removal, however, a change of caliper will be necessary to accommodate the extra thickness. The caliper was specific to the V8 but the Triumph 2000/2500 unit is very similar. The difference is the outer half which has an extra casting ridge which will foul standard MGB rims. This may not happen with non-standard rims but a solution is to use the Triumph inner with the original MGB outer to produce a hybrid. This rebuilding is best done by brake specialists. Pads from the Ford Transit (two pot caliper) are the same as, but cheaper and more widely available, than those for the B V8.

With the thicker disc comes the option of using the four piston (four pot) calipers from the Princess/Ambassador to produce a set-up not unlike that of the RV8. This gives a good increase in pad contact area – which is better controlled by the four pistons. A change of mounting bolts (high tensile) are required for use on a B, or re-thread the stub axle.

The pipework for the hydraulics presents a few extra problems as this caliper operates with two separate circuits, each with two pistons. If your MGB has single circuit hydraulics then the simplest option is to fit a T-piece into the feed line to provide two flexible supplies. Where the car has twin circuits you can increase safety margins by splitting the system more completely; for example, an 'H' 'I' split, where the 'H' feeds to two pistons on each of the calipers and to both rear cylinders on the primary circuit, and the 'I' has the secondary circuit feeding the remaining two pistons of each caliper. Thus, in the event of a circuit failure, the front, where most of the efficiency and stability comes from, always retains braking power. This is one example but there are other routes that can be followed.

Because of the popularity of this conversion amongst Ford tuners, a range of uprated pads are available, while standard Princess pads are readily available from local motor factors.

Vented discs are the last option for upgrading the front. They do not improve braking, unless of a larger diameter than those they replace, but do dissipate heat faster. Ideally, the vented disc should have a diameter very close to that of the original, so the braking remains unchanged whilst the venting provides the extra surface area to dissipate heat. There is no point in going smaller! These discs necessitate the use of alternative, wider calipers to span the thicker disc. Reconditioned Princess four pots are available with spacer kits to suit (intended for Ford conversions).

For road conversions be wary of using calipers which may have originally been developed for competition use. These may not have the same sealing arrangements to resist the effects of salt, mud, snow, etc on a continuous basis, which may lead to problems later. It makes more sense to use calipers from another road car, where such problems have been taken into account and service parts are more readily available.

One other consideration when using a caliper with a different number of, and/or different sizes of piston, has to be the ratio between the master cylinder, caliper and the other slave cylinders in the circuit. The conversion to the Princess calipers maintains this ratio perfectly for all the later dual circuit MGB cars. Single circuit MGBs use a master cylinder of slightly smaller bore which results in slightly longer pedal travel to achieve the same piston displacement at the wheels. Replacing the master cylinder with one of a larger bore will restore the status quo.

So far attention has focused on the front brakes. The rear drums are fine and provide reliable long-term service. However, an increase in efficiency at the front could cause the rear wheels to lock up under very hard braking (increased weight transfer forwards); an unstable condition which must be remedied. This is done by reducing the diameter of the rear wheel slave cylinders from 0.8in to 0.75in (20.32 to 19.05mm).

Alternative slave cylinders are available from Moss. If some other size should be required there is a choice from the vast Lockheed range. Each cylinder has a locating dowel to suit the intended application. This means either the backplate or the cylinder will require alteration to fit a B. When the correct size of cylinder has been fitted the standard linings will give the correct balance of braking effort.

Lastly, a quick mention of cooling. If your car has a deep front spoiler fitted it can impede the flow of cooling air to the brakes unless equipped with ducts for that purpose. Air inlet ducts are available from competition accessory suppliers. For competition, using large diameter flexible trunking to supply air directly to the back of the brakes is a good idea if the regulations allow.

WHEELS AND TYRES

This is quite an emotive subject with many people considering appearance to be a priority over effectiveness. It must first be decided whether anything other than standard components are to be used. If so, whether to use 14 or 15 inch diameter wheels, which are, to all intents and purposes, the two sizes to choose from. The decision can be influenced if a large diameter brake conversion is to be fitted. The standard rear bodywork presents the next restriction as ultra-wide wheels and tyres don't fit the available space, which is why the RV8 has to use a bulged rear wing design.

If the body is to remain standard then 6 inches is the maximum practical rim width, in either 14in. or 15in. diameter. A 185/70 tyre for the 14in. should operate without body contact. For the 15 inch wheel a 185/65 tyre is the largest that can be considered, and even then some modification to the lip of the wheelarch will probably be needed. Reducing the profile yet retaining the section width eases the problem, so a 185/60 x 15 tyre normally operates without contact. Most may then look at going to a 195/60, or a tyre with even lower profile.

A further consideration has to be the wheel offset, which dictates the relative position of the tyre within the arch. However, this problem has already been overcome for the common aftermarket alloy wheels available from MG suppliers. As long as you state the exact details of your vehicle the supplier will be able to provide a wheel and tyre for the job.

Tyres are another minefield. What is regarded by one person as the best tyre ever is rubbished by someone else! Tyre technology has come a long way since the B's heyday and can be readily applied to your car, unless you choose to pursue the original factory specification route. It is worth remembering that how well a car stops, handles and accelerates is dictated by the tyres. The choice of tyre manufacturer is a personal matter, be it for looks, performance or price.

It should be borne in mind that the suspension design of the MGB doesn't lend itself to the use of modern ultra-low profile tyres. Lower profile means less tyre sidewall deflection under load, giving increased feel and response but also greater transmission of the road imperfections to the car. For a B, very low profiles tend to make the steering over-sensitive to the point where handling is compromised. As a guide, 60 profile tyres should be regarded as the lowest to fit for balanced road performance.

Once wheels and tyres are chosen and fitted there is scope for altering the handling characteristics through tyre pressure variations. The bias between front and rear has a marked affect and can provide quite a degree of 'tuning.' There are handling advantages in increased pressures (to in excess of 40psi) but the ride suffers and tyre wear can increase. The loss of ride quality will usually dictate a stop long before these levels are achieved. Reducing pressure below the recommended levels for the car should not be done as this will give adverse results!

STEERING

When in good condition the standard steering rack is fine for all purposes. The only variation may be to fit a smaller diameter steering wheel of, say, 13 or 14 inches. This will make steering more responsive but also slightly heavier.

Chapter 14
Setting up & rolling road tuning

Although rolling road dynamometers have been in use since 1940, most car owners do not appreciate the performance benefits that can be gained from using one, even on a standard car. Contrary to a common misconception, the dyno isn't there just to 'screw the nuts' off the car and see what power it develops at valve bounce, or to give a graphic display of con rods exiting the block when you lift off the throttle at 7000rpm! It is used as a means of checking an engine for efficient running under controlled conditions.

The car can be driven at various throttle openings to enable sampling of the exhaust gases, and monitoring of the various electrical output signals to ensure that all is in order. In high-tech jargon it's 'real time monitoring' of the engine as it is working, and a weak mixture will show up rapidly, as will a sparkplug misfire. It allows diagnosis of faults that don't show up with usual 'static' tuning methods. Of course, for the dyno to be effective

it needs a skilled and knowledgeable operator, but a rolling road still takes the guesswork out of tuning. Any competent amateur armed with an exhaust gas analyser or air/fuel ratio sensor, a degree (dial-back) timing light and dwell/volt meter can achieve similar results, but only with the car stationary. The efficiency of the engine when actually working under load would still be a seat-of-the-pants diagnosis without a rolling road to verify the adjustments – unless the tuner could run alongside the car making adjustments whilst carrying all the diagnostic kit, monitoring the readings, and dodging the traffic!

So here's a step-by-step guide to dyno tuning a standard MGB (the basic procedures would be the same for a modified car). Once the car has been lined up on the rolling road, the front wheels are chocked to prevent it making a rapid exit from the rollers. Covers are then placed on the front wings to protect the paintwork and provide a

An MGB on a rolling road cruising in third gear at 3000rpm, 30mph and 17bhp at the wheels.

useful surface on which to put spanners, carburettor bits, screwdrivers, etc. When you are working under the bonnet you need the tools close to hand, especially with your back and one arm bent double when adjusting the mixtures on HS4 SUs, and needing to alter the idle rpm with the other hand!

What is probably the most important step comes next – checking oil and water levels. Engines don't function very well for long without them! Although it is fair to say that the majority of people check these levels in their cars as a matter of course in preparation for the tuning session, it never hurts to double-check anyway.

While the leads are being connected from the diagnostic analyser, a visual check is made around the engine bay for any devious little faults that may be lurking, such as leaky fuel lines, missing carburettor overflow pipes, cracked wiring looms that can earth out live wires, or the perennial B problem of water leaks on the plug side of the head, with the occasional head cracked between numbers two and three sparkplugs thrown in for variety. These faults will need swift rectification before the testing begins, except in the case of the cracked head which can be sorted out at some other time.

With the leads connected – earth to diagnostic, live feed from the car's fuse box, feed from both sides of the coil (+ and -), an inductive pick-up from the coil HT lead and another inductive pick-up from the number one HT lead – the electrical system can be checked.

The machine works by the inductive pick-up on the coil HT lead detecting leakage from the lead, and this is displayed as voltage against time on the oscilloscope. The pick-up on the number one lead triggers the oscilloscope to repeat this pattern – so you only see four spark lines (traces) for a four cylinder engine, eight for an eight cylinder unit and so on. The number one pick-up also registers engine rpm and produces the signal that fires off the timing light.

By far the most common electrical fault with a B is burnt-out points. The voltage across them tends to be too high and fluctuates whenever the points arc, which causes a high speed misfire.

If the condition of the points is okay, or if new points have been fitted, then the points gap is measured. This is most accurately checked using a dwell meter, which records an average reading of how long the points are closed (displayed on the diagnostic in degrees). Feeler gauges do not allow for the inevitable wear that you find on distributor shaft bearings. Having set the points to fourteen or fifteen thousandths of an inch with feeler gauges, the actual gap may be less when the engine is running, decreasing the dwell time and so reducing the coil's output and engine efficiency.

With the points functioning correctly the oscilloscope now shows any sparkplug, plug lead or distributor cap faults, which must be rectified before progressing to the next stage of checking the vacuum and centrifugal (mechanical) advance characteristics of the distributor. The timing at idle is checked with the vacuum advance pipe disconnected, and the total mechanical advance is checked by revving the engine to 4000 or so rpm. This is where the degree-type timing light is best, so you can dial back to align the Top Dead Centre (TDC) marks on the block and pulley, otherwise you need to accurately mark the 34 degrees Before Top Dead Centre (BTDC) position on the pulley itself and that's difficult. The vacuum pipe is reconnected afterwards.

A fairly common fault with high mileage distributors is tired centrifugal advance springs, which allow the distributor to advance too much at idle. With the timing set correctly at idle but with worn springs, the maximum advance would be too low. For example, if the distributor is normally set at 14 degrees BTDC at idle and the distributor gives 20 degrees of mechanical advance,

Setting SU carburettor jet heights.

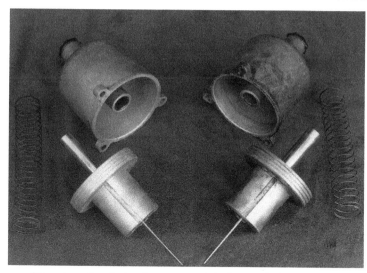

SU carburettor dashpot components before and after cleaning.

engine heat (it's brown in colour instead of bright clean aluminium). This must be removed, using choke cleaner or thinners, in order to allow the piston to slide freely inside the dashpot. Any dirt on the piston itself is also cleaned

then the overall maximum timing should be 34 degrees. However, if the springs are worn and allow the distributor to advance, say, 10 degrees the moment it turns, then the distributor can only advance the remaining 10 degrees. This would give us the 14 degrees of static advance (which would now include the extra 10 degrees), plus the remaining 10 degrees of mechanical advance, which equals a total of only 24 degrees maximum overall advance, – definitely not enough for a B!

Once the ignition system has been checked over and is seen to be in reasonable working order, the next stage is to set-up the carburettors.

The air filters need to be removed first as this gives better access to the carburettors. The dashpots are removed and the jet heights equalised. A good starting point is around 60 thousandths of an inch (1.5mm) down from the bridge. Ideally, the jets should already be within a thou or two of each other; any more than 10 thou difference and carburation will be upset. In most cases, when you look inside the carb dashpot you will see that the inside of the pot is encrusted with burnt oil and varnish from the

off at the same time. This accumulation of muck will, believe it or not, cause loss at the wheels of at least 3 to 5bhp!

Once the jet heights are set, the throttle linkages need checking to see if they are working in unison; if not, the result is one carb working more than the other, and cruising conditions will be poor.

The dashpots are refitted and filled with engine oil to the top of the rod – no oil will make the fuel mixture weak during acceleration, causing the engine to cough and splutter for a few seconds.

The next job is to check that airflow through each carburettor is the same. With the engine idling, use a carb balancer (or a short length of rubber tube) to listen to each carb in turn and equalise airflow using the idle screws. If necessary, the throttle linkage can be re-adjusted afterwards. With the engine switched off, a quick check is again made to see if full throttle is achieved by depressing the accelerator pedal fully, and testing the carb linkage for any further movement with the pedal still depressed. An assistant is useful here, unless you're very flexible and have very long arms!

The air filters are refitted and the car is ready to be tested on the rolling road. With the engine running, the idle mixture is set to give a nominal 3 to 3.5 per cent CO reading. The big cooling fan is switched on and, with the car running in third gear and showing 10 to 12bhp at the wheels at 30mph, the mixture is noted. A reading of somewhere between 0.75 to 2.5 per cent CO is reasonable for a standard B. The throttle is then gently depressed and the speed increased along with the bhp reading at the wheels. A further check is made with the car running on part throttle at 4000 to 5000rpm, which is fast cruising, and the mixture should still be around 0.75 to 2.5 per cent CO.

With the cruising mixtures checked, next is full throttle testing, beginning from about 3000rpm. This is achieved by loading the rolling road, which makes it more difficult for the car's wheels to turn the rollers – it can be likened to driving up a steep hill, but I can control the slope of the hill to restrict the engine speed. The mixture should now be reading around 5 per cent CO. If the engine seems OK and oil pressure and water temperature are good, then a full throttle 5500rpm mixture check is done. The mixture should be around the 5 per cent CO mark. At this stage the peak power output is noted – usually around 65bhp at the wheels for standard engines.

If the engine makes peak power at less than, say, 4800rpm, then the cam or the rockers may be worn, reducing valve lift. When the engine is further loaded down to 3000rpm the power is again noted and should be around 42 to 44bhp. If the car shows more power here, then the cam or rockers are worn – losing valve lift makes the cam behave in a more mild-mannered way than usual, with more mid-range power and less at high revs.

The final stage is to load the engine down to 1500 to 2000rpm to see if it pinks on full throttle or part throttle. If it pinks on full throttle then the distributor must be retarded until pinking stops. If the engine only pinks on part throttle and not on full throttle, there is too much vacuum advance – remember, vacuum advance only works on light throttle openings. If it's very bad then the distributor will need a replacement vacuum advance canister. If the engine doesn't pink then the rolling road session is complete and the car will now be smoother to drive, have sharper throttle response and generally perform in a manner guaranteed to put a smile on your face.

MODERN DYNO TUNING SESSION

In this section we are able to describe a typical rolling road tuning session step-by-step, showing the results as we go along. In this example the work is performed on a chrome bumper MGB GT belonging to Simon McCloud.

Upon the car arriving at the unit we do a quick visual inspection, just to get an overall 'feeling' for the condition; tyres, bodywork, exhaust, etc. The next step is to drive the car onto the rolling road, get it positioned correctly, and then align it straight in the rollers by steadily driving it in first gear. Then we can also see if the car sits straight in terms of suspension and body to axle alignment. Any perceived problems and we ask the customer about the history of the car. Simon's car was obviously well cared for and beautifully turned out, so there were no issues of note.

The tyre pressures of the driving wheels were checked and set at 30psi for testing purposes. This removes tyre pressure as a testing variable and maintains consistency for any subsequent dyno testing at a later date.

Our new rolling road measures power at the wheels by either accelerating a known mass (the weight

Engine bay of the chrome bumper MGB GT used on the dyno.

of the two rollers) in a known time, which gives horsepower, or by using a power absorption unit (PAU) to hold full throttle loads. It can also use a combination of both inertia and PAU. The acceleration tests are very gentle on the engine and typically only take about seven seconds per test. Sensors and data gathering equipment on the dyno combine with sophisticated computer software to produce graphical power curves from the tests. The results can be corrected to various test standards but we use SAE J1349.

The car is carefully strapped down to guarantee good tyre contact with the rollers so power output can be measured consistently; some people worry that strapping down is to stop the car flying out of the rollers, but this is not the case.

After a quick under-bonnet visual inspection we perform four or five unmeasured acceleration runs to let everything settle in, after

which the tension of the anchoring straps is re-checked. The warming up process also incorporates all the bearings on the rolling road, again to ensure measurement consistency and repeatability. If the engine and transmission are cold the car is warmed up by being driven on the rollers for five minutes or so.

We usually run the cars by first 'telling' the data gathering software what a specific rpm is in terms of mph in a specific gear (usually 2000rpm and 4th gear).

The software can lock in this figure so all power runs compare exactly from that baseline for rpm and mph. We then perform an initial power run. When the system has settled again we do a repeat power run, this time starting from a slightly lower rpm. This produces a more realistic power curve with the engine now driving through the start point, and generates the baseline information needed so we can now test differences in tuning. If the power curves created at points above, say, 3000rpm overlay each other, we know we have steady good readings for repeatability.

We record the air/fuel (A/F) ratio by inserting a lambda sensor into the exhaust tailpipe; the same type sensor used by modern cars to monitor fuelling. An engine performs best with sufficient fuel for its needs. Too much or too little fuel will compromise performance.

The A/F sensor data can be plotted with the power curve, and from the graphs we can see if the engine is running properly in terms of fuel mixtures, or if it is too lean, too rich, or a combination of lean and rich. The cruising mixtures are also checked at this stage.

On Simon's car we observed the engine was fuelling a little on the rich side. The dashpots were removed from the 1½ inch SU carburettors and the needles checked to see if were the same, AAA in this case as the car was fitted with K&N air-filters, and likewise for both jet heights. The jet heights were adjusted by about 20 thou to weaken the mixture.

On re-test we were pleased to see better fuelling (less fuel used equals more economy) and a little more power at all rpm – a win-win situation. This can be seen in **Graph 1**.

The next task was optimising the ignition timing. Simon's car was fitted with an electronic 123 pre-set MG distributor. Of the 16 available pre-set curves provided, there are two curves in particular we favoured for this specification of engine, one of which we usually find best suits the actual state of tune. As the modified cylinder head fitted was an Econotune, with airflow developed towards producing an optimum power spread at lower rpm's, and a standard cam was fitted, we chose curve C, which has a fairly long advance

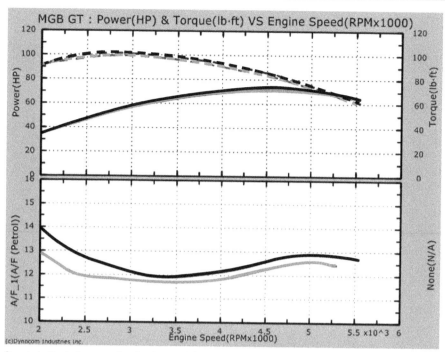

Graph 1: First step was to check the air/fuel and correct the slightly rich condition. The pale line shows the initial rich mixture condition and the darker line the result of the adjustments.

Checking the ignition timing.

curve. The ignition timing was checked on the car and initially set to give 32 degrees total advance at 4500rpm, the greater advance at high rpm helps to maintain horsepower when the engine is becoming a little 'breathless.'

After testing we set the timing to give 34 degrees total advance at 4500rpm, producing optimum power throughout the rev range. Power was gained at all rpm, as can be seen in **Graph 2**.

As part of the testing process we tend to first advance a distributor from the initial setting and take a power reading. If horsepower increases we try a little more advance and repeat until we get the best power without pinking/detonation. If extra advance loses power we try retarding the distributor. Sometimes we find power may improve at high rpm whilst reducing at low rpm, or vice versa. This lets us 'know' the advance curve is not correct. The curve may be too long (slow to reach max advance) or too short (quick to reach max advance) to suit the characteristics of the engine. With a car using a mechanical distributor with a fixed advance curve we can recommend the customer purchase a replacement distributor offering a better advance curve, or an electronic 123 preset version where we can select the advance curve to suit. If an electronic 123Tune distributor is fitted the shape of the timing curve can really be optimised by connecting a laptop. This could mean advancing or retarding the timing at specific discreet points; additions or subtractions not achievable with linear advance curve distributors. See also the Ignition Section, Chapter 9, Electronic Systems.

The third stage of tuning was to adjust the tappets to see if a little more power could be squeezed from the engine. Simon's car had a standard camshaft with tappet clearances 15 thou cold or 13 thou hot.

Previous testing has proven that the B series benefits from a wider hot clearance for the exhaust valves, around 15 thou, whereas the inlet valves favour

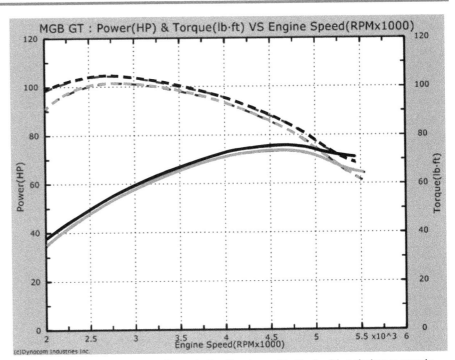

Graph 2: With the mixture adjusted attention turned to the ignition timing; a couple more degrees advance proving beneficial. The pale line is with the 123 distributor ignition timing set to 32 degrees total advance at 4500rpm, and the upper line is with the timing at 34 degrees at 4500rpm.

Resetting the tappets to our preferred clearances.

tighter hot settings of around 12 thou. Simon's B was no exception, and we gained a little power above 2300rpm. We lost a little below that due to the increased valve duration and overlap created by the tighter inlet clearances. Above 2300rpm win, below 2300rpm lose; as shown in the **Graph 3**.

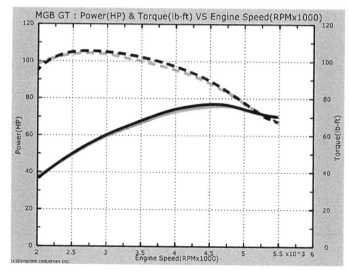

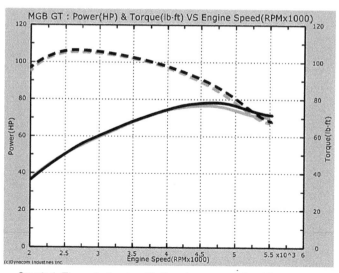

Graph 3: Mixture and timing done, time for things mechanical. The pale line is with the as arrived tappet clearances. The dark line is with the reset hot clearances of 12 thou inlet and 15 thou exhaust.

Graph 4: The pale line is with the plugs gapped to 28 thou, the dark line is the plugs cleaned and re-gapped to 35 thou.

Stage four, and we decided to clean and gap the sparkplugs. They were fairly clean, but were given a gentle grit blast, blown out with an airline to get rid of dust, and the gaps increased from 28 to 35 thou, which we find works best with UK unleaded fuels. There was a slight improvement in performance throughout the rev range, it being most marked above 4250rpm – another beneficial outcome. The accompanying **Graph 4** clearly shows this.

To recap, by adjusting the fuelling, ignition timing, tappet clearances and sparkplug gaps the max power at the wheels increased from 71.6 to 78.3bhp.

The transmission losses were measured as 22.6 horsepower, giving a flywheel output of 100.9bhp.

The final graph, **Graph 5**, shows how all the small changes added up to give a generous overall power improvement.

In addition to the measured power gain the car exhibited a sharper throttle response, a greater eagerness to rev and delivered the added bonus (reported to us later) of increased miles per gallon.

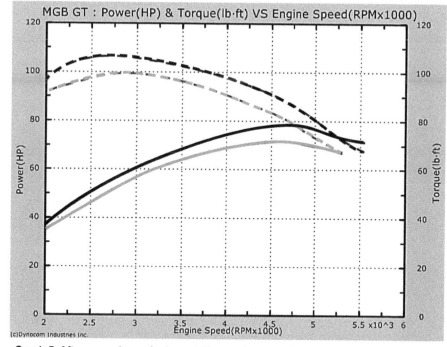

Graph 5: After a good morning's work the results speak for themselves; the pale line is the power at the wheels as arrived, the dark line is the result after tuning.

Of interest, based on previous experience, our target was 80bhp at the wheels. While the end result was nicely in the ballpark, Simon's car ran 85psi oil pressure and a mechanical cooling fan, which we know from experience costs two to three horsepower (see the graph in Chapter 6 – the oil pump).

Chapter 15

Improving & maintaining fuel economy

While it may seem strange, modifying a car for better performance and achieving better fuel economy are not contradictory goals.

Up to about 30mph the main resistance to the forward motion of a vehicle is from internal mechanical losses and friction losses from the tyres' contact with the road. Above that speed, rolling resistance remains but wind resistance (aerodynamic drag) comes increasingly to the fore, rising exponentially until the power needed to overcome it matches the remaining power available and maximum speed for the vehicle is achieved.

Driving techniques play a large part in achieving good miles per gallon. Things like minimum use of choke, smooth acceleration and deceleration, and driving in a higher gear when possible have a beneficial effect on fuel use.

The importance of setting up the car properly cannot be over-emphasised. Brakes, tyres, steering, suspension and the state of engine tune all have to be in good order to achieve the best mpg.

If the brakes are binding, you will lose power through increased frictional losses. It will require greater throttle opening to overcome the drag and achieve the same speed as with free running brakes. Obviously, this will use more fuel as well as wear out pads and linings rapidly. If the brakes are spongy, have air in the system, or pull to one side, you will have to drive to these conditions, and in compensating be less smooth, which leads to poorer fuel economy.

Obviously, steering and suspension influence how the car travels down the road. Worn suspension and bushes mean wheel alignment can be out; undue flex and give leaving the tyres pointing anywhere but straight ahead. If the car wallows and handles poorly you will not be able to employ a smooth driving technique.

Correct tracking (alignment) is important, both to lessen mechanical drag from the tyres on the road, and reduce tyre wear. Likewise, correct tyre pressures.

The detrimental effect of low tyre pressure can be seen, as shown in the graph. Here dropping the tyre pressures from 32psi to 18psi lost 5.7bhp at the wheels. A somewhat extreme example, but a very interesting experiment nonetheless!

Low tyre pressures increase road to tyre friction, creating more rolling resistance. To maintain the same speed as with correctly inflated tyres, more engine power is required, and more power means more fuel used.

A session on the rolling road to optimise fuelling and ignition timing should also include ensuring the sparkplugs, plug leads, distributor cap and rotor arm are in good condition – things which you can also do yourself during routine maintenance.

Even a failing exhaust system can be diagnosed – maybe the silencer innards are starting to collapse and

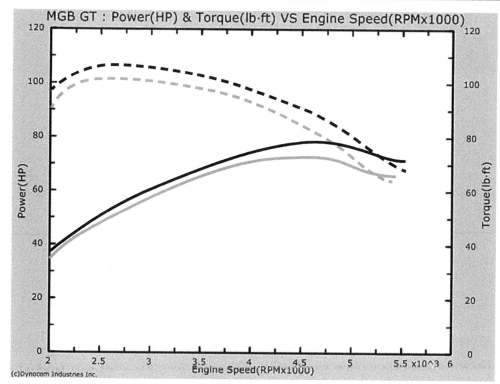

Graph: MGB tyre pressures.

hinder exhaust gas flow, or have disappeared altogether (tapping the silencer with a screwdriver handle should give a solid, flat sounding, clump; a clang means the wadding has gone), costing horsepower.

Say it takes 40bhp at 35% throttle opening to maintain a steady cruise speed of 60mph with a standard car. With a modified engine, or even a standard one that has been serviced/tuned, the improved efficiency means it could be making the necessary 40bhp with less throttle opening, using less fuel.

Depending on the type of rolling road used, the transmission losses could be measured and compared with those from an average car. We have seen extra thick axle oil, one containing special additives to 'cure' leaks, lose 3bhp over standard rear end oils.

The engine can be modified to improve efficiency. Fitting free flowing

air filters can reduce how hard the engine has to pull to draw in the air fuel mix. Improved cylinder head ports reduce pumping losses by getting the mixture in and waste gases out more easily. Better shape combustion chambers can help both flow and mixture motion, and when combined with a compression ratio increase, act to burn the inducted mixture more effectively: more energy is extracted from every combustion event.

A free flowing exhaust system helps expel waste gases more efficiently, as a further contribution to improving volumetric efficiency.

When it comes to the cylinder head, a combination of higher compression ratio and improved gas flow can pay dividends in terms of fuel economy. We initially developed the 'Econotune' specification cylinder head for a customer wishing to tow a caravan.

The requirement was to make the engine 'pull' better on hills without having to change down a gear. As building a larger capacity engine to increase torque was not an option, we only had the head to work with. We decided to keep the ports standard, apart from removing any 'bad' casting frazes. For the inlet ports, the coarse surface finish and the higher gas velocity created by retaining a smaller diameter helps fuel stay in suspension at lower rpm and smaller throttle openings. Rough finished exhaust ports encourage adhesion of combustion deposits that act as a thermal barrier and keep some of the heat from passing into the head. We decided to use the smaller 1.56 inch inlet valves and standard 1.34 inch exhausts. The smaller inlet valve keeps gas speed high and works well to around 4800rpm. The valve seats are cut with three angles (30/45/60), with a 60 thou (1.5mm) wide 45 degree seat.

The width of the 60 degree bottom cut into the port throat is reduced to approximately 60 thou (1.5mm) and blended into the opened-up valve throat. The short side of the throat is modified to leave a good radius.

The valves are modified on the port side with a 30 degree back cut, which is blended in to the 45 degree seat and back of the valve.

The result could be likened to a 'professional' DIY specification head.

The combustion chambers are smoothed but not opened-up on 18v heads. With 906/1326 head castings, the peak opposite the plugs can be relieved by around 1/8th inch (3mm). We tend to leave the chamber wall vertical with the peak dressed back.

An increase in compression ratio to approximately 9.75:1 improves the efficiency of the engine, too.

We have had excellent customer feedback regarding performance, fuel economy and our modified heads – the Econotune in particular. For example, when using an Econotune head and mild cam, or an Econotune head and standard cam, or even a Fast Road head and mild cam combination, 33/34mpg could be achieved when driving reasonably compared to 27/28mpg as standard. The Fast Road specification engine is a little less economical in overall driving conditions, but pretty well matched in cruising conditions.

Above right: The standard inlet valve (right) has a ridge behind the seat, which is an obstruction to smooth airflow. This is removed by adding a 30 degree backcut.

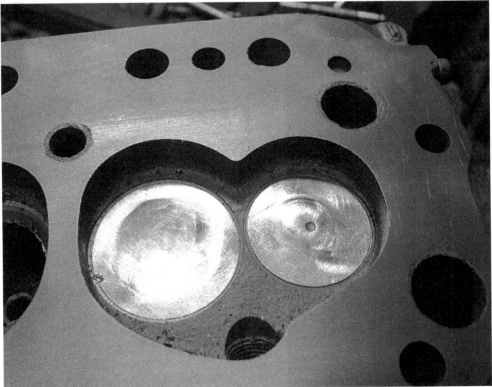

Standard small valve head combustion chamber.

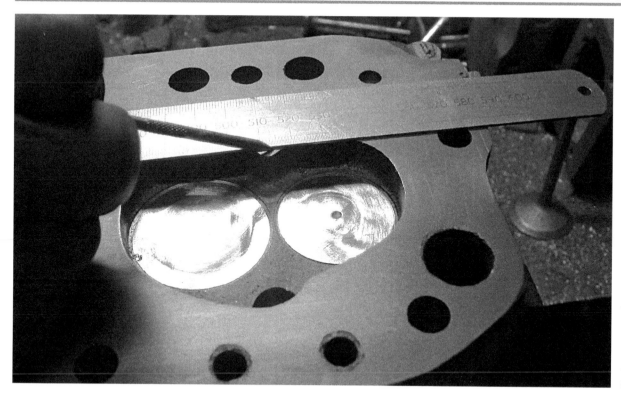

A line is scribed across the combustion chambers as a guide ...

... for grinding the peaks back square and vertical ...

... before final smoothing and blending.

Appendix 1
Useful formulae, data & cam timing

CALCULATING COMPRESSION RATIO

The compression ratio (CR) of an engine is arrived at by using the following formula:

CR = swept volume + clearance volume
          clearance volume

Where: swept volume (Vs) is the capacity of one cylinder in cm^3.

Clearance volume (Vc) comprises the cm^3 above the top piston ring, any dish or valve cutouts in the piston, the gap between the top of the piston at TDC and the top of the block, the compressed head gasket volume and the combustion chamber volume.
eg: $1798cm^3$ block, $6.5cm^3$ dished piston, $2cm^3$ above the piston, $4.5cm^3$ gasket volume and $43cm^3$ combustion chamber volume in the head.

Swept volume of one cylinder = 1798 = 449.5cm³
                                                  4

Clearance volume (Vc) = 6.5 + 2 + 4.5 + 43 = 56cm³
        CR. = 449.5 + 56
                56   = 9.03:1 (9.03 to 1)

If the CR needs to be raised the new clearance volume required can be calculated as follows:

CR - 1 = swept volume
            clearance volume

Therefore: *clearance volume = swept volume*
                                         CR - 1

eg: The new CR required is 10.5:1, what chamber volume is needed to achieve this?

Schematic diagram of four stroke engine for calculating CR.

Using the same cylinder volume as previously (449.5 cm³):

clearance volume = $\dfrac{449.5}{10.5 - 1}$

$\dfrac{449.5}{9.5}$

= 47.32

As we know the clearance volume values as a 6.5cm³ dish in the piston, 2cm³ above the piston and 4.5cm³ gasket volume, the chamber volume can be established by subtracting these values from the calculated clearance volume; eg:

New chamber volume = 47.32 - 6.5 - 2 - 4.5 = 34.32cm³

CALCULATING HOW MUCH TO SKIM OFF THE HEAD

New chamber depth required = $\dfrac{\text{original depth x required } cm^3}{\text{original } cm^3}$

eg: cm³ required is 34.32, original is 43cm³ and the original chamber depth is 10mm.

New chamber depth = $\dfrac{10 \times 34.32}{43}$

= 7.98mm

Therefore skim 10 - 7.8 = 2.02mm (79 thou) from the head face.

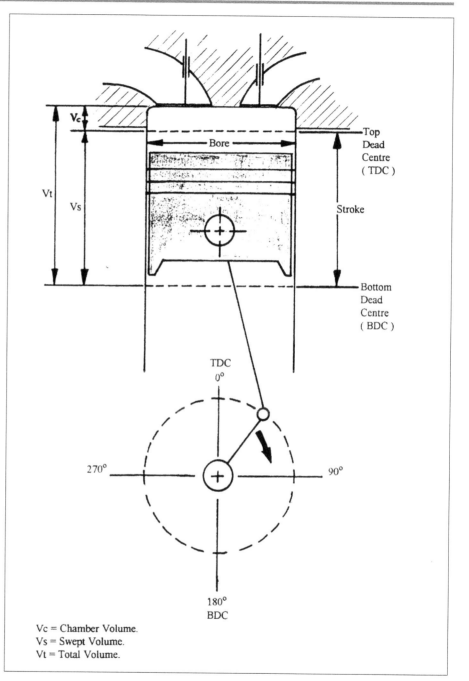

Vc = Chamber Volume.
Vs = Swept Volume.
Vt = Total Volume.

GAS SPEED THROUGH INLET VALVES - EFFECTS OF ENGINE SIZE AND CAMSHAFT SELECTION

Peak power rpm = $\dfrac{\text{gas speed x 5900 x valve area}}{\text{cylinder volume}}$

Where: gas speed is in feet/second; valve area is in square inches; cylinder volume is in cm^3

For a standard large inlet valve 1798cc MGB, peak power is at 5000rpm, valve area is 2.074^3 and cylinder volume is $449.5cm^3$.

Gas speed = $\dfrac{rpm \times cylinder\ volume}{5900 \times valve\ area}$ = $\dfrac{5000 \times 449.5}{5900 \times 2.074}$ = 183.7 ft/sec.

With a standard cam and a 1950cc engine ($487.5cm^3$ per cylinder) the engine will peak at a lower rpm as the gas speed will be reached earlier.

Peak rpm = $\dfrac{183.7 \times 5900 \times 2.074}{487.5}$ = 4611rpm (Close to the 4700rpm peak found on the rolling road).

If the 1950cc engine is fitted with a tuned cam that increases maximum gas speed to 195 ft/sec. -

Peak rpm = $\dfrac{195 \times 5900 \times 2.074}{487.5}$ = 4895rpm (Close to the 5000rpm peak found on the rolling road).

TORQUE AND HORSEPOWER

The relationship between torque and horsepower:

Bhp = $\dfrac{rpm \times torque}{5252}$

Torque = $\dfrac{bhp \times 5252}{rpm}$

Torque figures are measured directly from the engine by a dynamometer and then mathematically converted to horsepower (bhp). For example, if an MGB engine produces 79bhp @ 3000rpm, 107bhp @ 4000rpm and 128bhp @ 5200rpm. How does the Torque output vary?

Torque = $\dfrac{79 \times 5252}{3000}$ = 138.3lb/ft

Torque = $\dfrac{107 \times 5252}{4000}$ = 140.5lb/ft

Torque = $\dfrac{128 \times 5252}{5000}$ = 129.3lb/ft

From this can be seen that the torque output curve is fairly flat – a good torquey (lots of grunt) engine. These figures are from a 1900 MGB with a fairly fast road cam. At 5252rpm bhp and torque must be equal. Torque produces acceleration, whereas bhp provides top speed. Ideally, the rpm difference between peak torque and peak horsepower wants to be as large as possible (then the engine will pull like a steam train and still rev well!) providing maximum flexibility.

TORQUE AND ACCELERATION

If the total weight of the car and engine maximum torque are known then 0 to 60 acceleration can be estimated using the following formula:

0 to 60 (seconds) = $\dfrac{2 \times W \text{ }^{\wedge}0.6}{T}$

Where: W = weight of car T = maximum torque in lb/ft

Don't forget to include the driver's weight and any fuel in the tank (approx 7.5lb per imperial gallon)

eg: A standard MGB GT weighs 2395.8lb, plus driver (say, 150lb): max. torque is 100lb/ft.

0 to 60 = $\dfrac{2 \times 2545.8 \text{ }^{\wedge}0.6}{100}$ = 10.6 seconds.

Increasing torque will reduce the 0 to 60 time. For instance, the engine used as the example for the torque and horsepower maths produced a maximum torque of 145lb/ft.

0 to 60 = $\dfrac{2 \times 2545.8 \text{ }^{\wedge}0.6}{145}$ = 8.46 seconds.

Suppose the car weight is reduced by, say, 200lb. What would be the effect on the 0 to 60 time?

0 to 60 = $\dfrac{2 \times 2345.8 \text{ }^{\wedge}0.6}{145}$ = 8.05 seconds.

If maximum bhp is known, a rule of thumb method of establishing the increase in top speed is as follows:

Max. speed = $3\sqrt{\dfrac{\text{new bhp}}{\text{old bhp}}}$ x old max. speed

eg: A standard MGB GT has 65bhp at the wheels and a top speed of 108mph. With 90bhp at the wheels:

Max. speed = $^3\sqrt{(90/65)}$ x 108   = $^3\sqrt{(1.385 \times 108)}$ = 1.115 x 108   = 120.4mph

eg: How much bhp at the wheels is necessary to reach 145mph?

New bhp @ wheels = $\dfrac{\text{new mph}}{\text{old mph}}^3$ x old bhp   = $\dfrac{145}{108}^3$ x 65   = 2.42 x 65   = 157bhp at the wheels! (you would need a V8 for this!)

FUEL SUPPLY

The following formula can be used to arrive at an approximate idea of the engine fuel supply needs in order to be able to size the fuel pump correctly.

Fuel consumption (lb per hour) = max. engine horsepower x BSFC

BSFC stands for Brake Specific Fuel Consumption, which is given in pounds of fuel used per horsepower per hour. Most engines run figures of between 0.45 (very efficient) and 0.6 (thirsty) so 0.53 is a reasonable value for a roadgoing B.

So, with 1 Imperial (UK) gallon of fuel weighing around 7.5lb:

gallons per hour = (max. engine hp x 0.53) divided by 7.5

eg for a race car (170hp x 0.6) divided by 7.5 = 13.6 gallons per hour, so select a pump that can supply above that value – typically 18 Imperial gallons per hour. (To convert Imp gallons to US gallons multiply by 1.20095)

UNDERSTANDING CAMSHAFT SPECIFICATIONS

Most manufacturers supply the camshaft specification as a table of numbers, of which the timing figures are the ones we can work with; eg: cam timing 16/56 51/21

This means the inlet valve starts to open at 16 degrees before Top Dead Centre (TDC) and closes 56 degrees after Bottom Dead Centre (BDC). The exhaust valve starts to open 51 degrees before BDC and closes 21 degrees after TDC.

The overlap period where both inlet and exhaust are open together is calculated by adding the first (inlet valve opens) and the last (exhaust valve closes) figures together: 16/56 51/21 - 16 + 21 = 37 degrees. So this cam has 37 degrees overlap.

The duration (the time the valves are open) is calculated by: 16 + 56 + 180 = 252 degrees duration for the inlet and 51 + 21 + 180 = 252 degrees duration for the exhaust.

The 180 degrees is one half of a stroke that each induction or exhaust cycle takes.

The lobe centre angle (point of peak cam lift) for the cam is calculated by:

$$\frac{Duration}{2} - valve\ opening\ point\ for\ the\ inlet\ or\ closing\ point\ for\ the\ exhaust = \frac{252}{2} - 16 = 110\ degrees\ for\ the\ inlet$$

$$or = \frac{252}{2} - 21 = 105\ degrees\ for\ the\ exhaust$$

The inlet full lift point provides a starting point for cam installation if none is specified by the manufacturer. Otherwise you may find these figures vary from the manufacturer's recommended installation figures. ie, the inlet peak cam lift may mathematically work out at 110 degrees (after TDC) but the recommended installation is 108 degrees, two degrees retarded (the inlet valve reaches peak lift sooner). This is usually because during testing the cam was found to work best timed in at this figure.

The maths is only right for asymmetric cams with the same opening and closing rates on the lobe. Non-asymmetric cams can have peak cam lift actually occurring either side of this figure, depending on the lobe profile. In this case it is best to follow the manufacturer's recommendations.

The lobe separation angle (LSA) is calculated by:

$$\frac{inlet\ lobe\ centre\ angle + exhaust\ lobe\ centre\ angle}{2} = \frac{105 + 110}{2} = 107.5\ degrees$$

Of course, all these figures are only useful as guidelines due to the differences in advertised timing figures used by various manufacturers; it depends what valve lift, if any, they quote the figures from.

SIMPLE GEAR CALCULATIONS

3 syncro gearbox, standard ratio. Rev drop 1st to 2nd.

$$\frac{3.6363}{2.2143} = 1.642$$

5500rpm in 1st drops to: $\frac{5500}{1.642} = 3349rpm$

3 syncro gearbox, close ratio. Rev drop 1st to 2nd.

$$\frac{2.450}{1.620} = 1.512$$

5500rpm in 2nd drops to: $\frac{5500}{1.512} = 3636rpm$

Rev drop into O.D. 3rd

5500rpm in 3rd drops to: *5500 x 0.802 = 4411rpm*

It is possible to calculate speed, if the overall tyre diameter (loaded) is known (measure tyre's vertical height on car for approximation), using the formula:

Speed = overall tyre diameter (inches) x engine rpm
$\qquad$ *Differential ratio x gear ratio x 336*

On most cars top gear, 4th, is 1.00. i.e:
$$\frac{23.5 \text{ inches} \times 1000\text{rpm}}{3.909 \times 1.00 \times 336} = 17.9\text{mph}$$

To include the overdrive ratios multiply the gear ratio by the overdrive ratio (for example, third 1.3736 x 0.802 = 1.101) and use this in the bottom line (in above equation it gives 16.25 mph/1000).

If the overall tyre diameter is unknown but the figure for tyre revolutions per mile is, then the tyre diameter can be calculated by:

Tyre diameter (inches) = $\frac{20800}{\text{revs per mile}}$

Or revs per mile from overall diameter

Revs per mile = $\frac{20800}{\text{overall tyre diameter}}$

Changing the tyre size can have a marked effect on overall gearing. The change can be calculated by using the revs per mile figures for the old and new tyres.

$\frac{\text{Revs per mile (new tyre)}}{\text{Revs per mile (old tyre)}}$ x final drive ratio = effective new final drive ratio.

In effect, changing tyres can be likened to changing the differential ratio in the axle.
Say you change from a 165 x 14 tyre, with a revs per mile figure of 854 to a 185 x 15 with a revs per mile figure of 794, with the original 3.9 differential. Putting the numbers into the equation gives an effective new final drive ratio of 3.6. This will change O.D. top from 22.3mph per 1000rpm to 24mph per 1000rpm, reducing the engine speed for a given miles per hour figure – ie more relaxed motorway cruising.
The formula for a change of differential to preserve the original ratio is:
$\frac{\text{Revs per mile (old tyre)}}{\text{Revs per mile (new tyre)}}$ x final drive ratio = final drive to preserve same overall gearing

Which, in the previous case, means using a 4.2 final drive. Of course, in many cases the required ratio will not be available, so the nearest available to it will have to be selected.
The speedometer reading will also be affected by a change of tyre and/or rim size, the error can be calculated by:

$\frac{\text{Revs per mile (old tyre)}}{\text{Revs per mile (new tyre)}}$ x indicated speed = actual speed

The revs per mile figure can be replaced by the overall (loaded) tyre diameter measurement, old and new, in the above equation. The alternative is to check the readings on a calibrated rolling road; ask when you next take your car.

SU NEEDLE SELECTION CHART

Application	HS4 fixed	HS4 biased	H1F4 biased
Std	5	AAU	ACD
K&N modified	6	ABD/AAA	AAA

Application	HS6 fixed	HS6 biased	H1F6 biased
Std	KP	BBW	BBW
K&N modified	TE	BDR	BDR

JET SETTINGS SIDEDRAUGHT WEBER & DELLORTO

	Weber 45 DCOE			Dellorto 45 DHLA		48 DHLA Split twins	
	Std	F/road	Sprint	Std	F/road	F/road	Race
Choke	36	38	40	35	35	38	42
Emulsion tube	F16	F16	F16	5	8	5	5
Main jet	170	185	195	175	185	180	190
Air jet	155	160	120	160	165	160	145
Pump jet	45	45	50	55	55	50	60
Idle jet	50F8	50F8	60F8	50	50	60	60

CAM SPECIFICATIONS
Piper cams

	Rpm	Valve timing		Duration		Valve lift		Valve clearance		Inlet full lift		
		In	Ex	In	Ex	In	Ex	In	Ex	ATDC	LSA	Overlap
B851 Ultimate race	3500-7800	58-90	88-60	328	328	0.536	0.534	0.016	0.018	106	105	118
B929 Race	2500-7000	44-76	76-44	300	300	0.360	0.360	0.014	0.016	106	106	88
BBP255 Mild road	1000-6000	27-63	64-28	270	270	0.405	0.389	0.012	0.014	108	108	55
BBP270 Fast road	1500-6500	31-65	65-31	276	276	0.405	0.403	0.014	0.016	107	107	62
BBP285 Ultimate road	2000-6750	32-66	66-32	278	278	0.445	0.445	0.014	0.014	107	107	64
BBP300 Rally	2500-7000	49-81	81-49	310	310	0.460	0.458	0.016	0.018	106	106	98
BBP320 Race	3000-7500	53-79	79-53	312	312	0.465	0.465	0.014	0.014	103	103	106
HR270 Fast road	2000-6500	28-64	64-28	272	272	0.397	0.397	0.016	0.018	108	108	56
HR270/2 Fast road	1800-6500	22-70	62-28	272	270	0.400	0.400	0.012	0.014	112	110.5	50
HR285 Road/Rally	2500-7000	37-71	71-37	288	288	0.405	0.405	0.015	0.016	107	107	74
HR285/2 Road/Rally	2200-7000	35-77	63-35	292	278	0.437	0.437	0.014	0.014	109	107.5	70
HR300 Rally	3500-7500	42-74	74-42	296	296	0.450	0.450	0.014	0.016	106	106	84
HR300/2 Rally	3250-8000	46-74	78-42	300	300	0.459	0.459	0.015	0.015	104	106	88
HR320 Race	4500-8000	54-82	82-54	316	316	0.440	0.440	0.017	0.018	104	104	108
HR320/2 Race	4500-8500	60-88	88-60	328	328	0.470	0.470	0.018	0.018	102	104	120
HR330 Race	5000-9000	54-82	97-63	316	340	0.462	0.462	0.017	0.018	102	105.5	117

BLMC Special Tuning Profiles

| | Rpm | Valve timing | | Duration | Valve lift | Valve clearance | | Inlet full lift | | |
		In	Ex			In	Ex	ATDC	LSA	Overlap
Original MGA & MGB	-	16-56	51-21	252	0.361	0.015	0.015	110	107.5	37
714 Fast road	2000-6000	24-64	59-29	268	0.358	0.015	0.015	110	107.5	53
864 Road/Rally	2500-6500	36-52	56-32	268	0.358	0.018	0.018	98	98	68
770 Race	3000-7000	50-70	75-45	300	0.445	0.018	0.018	100	102.5	95
862 Race/Sprint	3500-7500	60-80	75-45	320/300	0.445	0.018	0.018	100	102.5	105
863 Race/Supersprint	4000-8000	60-80	85-55	320	0.445	0.018	0.018	100	102.5	115

Kent cams

| | Rpm | Valve timing | | Duration | | Valve lift | | Valve clearance | | Inlet full lift | | |
		In	Ex	In	Ex	In	Ex	In	Ex	ATDC	LSA	Overlap
714 Mild road	1800-5800	25-65	60-30	270	270	0.358	0.358	0.015	0.016	110	107.5	55
715 Sports	2000-6000	37-69	69-37	286	286	0.358	0.358	0.016	0.016	106	106	74
716 Supersports	2500-6500	46-76	76-46	302	302	0.375	0.375	0.018	0.020	105	105	92
717 Sports	1800-6500	37-65	73-27	280	280	0.394	0.394	0.022	0.024	103	108.5	64
718 Supersports	2500-7000	42-68	78-32	290	290	0.416	0.416	0.022	0.024	103	108	74
719 Race	3000-7500	47-73	83-37	300	300	0.438	0.438	0.022	0.024	103	108	84
Scatter pattern cams (altered lobe separation angles & phasing (full lift timing) for different cylinders)												
717SP Sports	1800-5600	37-65	65-35	280	280	0.394	0.394	0.022	0.024	103	104	72
718SP Supersports	2500-7000	42-68	70-40	290	290	0.412	0.412	0.022	0.024	103	104	82
719SP Race	3500-7500	47-73	75-45	300	300	0.438	0.438	0.022	0.024	103	104	92
720SP Race	4000-7500	46-74	80-48	300	308	0.436	0.446	0.022	0.024	104	105	94
721SP Race	4000-8000	50-78	81-49	308	310	0.448	0.455	0.022	0.024	104	106	99

Kent valve lifts are quoted using a theoretical rocker ratio of 1.42:1. Duration is quoted at 0.016in checking height.

CAM TIMING

The correct installation and timing in of the camshaft is, without a shadow of a doubt, one of the most vital steps toward achieving a powerful, smooth and progressive engine.

You may have been led to believe that it's just a matter of fitting the camshaft timing gear and aligning the dots, just as it says in the manual,

right? Wrong! In the majority of cases tolerances of various components can add up to the camshaft being a considerable way off the manufacturer's intended setting. This particularly applies to reground aftermarket profiles, especially now that original BMC cams are no longer available, where, in order to reprofile a worn out original, the cam timing can become changed due to the

nature of the reprofiling work. Also, the correct installation position can be very different from standard if an aftermarket performance cam is being fitted; the recommended installation position is usually given on the specification sheet supplied with the cam.

As the camshaft is really the 'brain' of the engine in that it controls opening and closing of the valves and the time of opening and closing during the four stroke cycle, any significant variation from this intended position can considerably alter engine power delivery characteristics. Small variations in timing may have no affect at all on engine performance, but you can never tell. So, correct timing in of the cam really is essential if disappointment in engine performance is to be avoided.

There are various methods for timing in a cam and the most common for a cam in block engine is by using the inlet full lift position, which is now described.

To correctly time in the camshaft requires a few pieces of equipment. These comprise a dial gauge, on a magnetic stand for preference, a large 360 degree protractor (called the timing disc or degree wheel – available from cam companies) and a short length of wire that can be bent to form a pointer.

With the camshaft installed, fit the cam gear and timing chain – there's no need to fit the camshaft retaining nut and washer yet – and align the dots on the two sprockets, all as given in the relevant workshop manual.

The timing disc can then be fitted to the front of the crank using the bolt for the front pulley. This will need a couple of suitable thick washers or somesuch as packing, fitted behind the disc, otherwise the bolt will bottom in the crank before it clamps the disc. Very lightly nip up the bolt with your fingers, as the disc will need to be rotated in order to zero it later. The wire pointer should then be attached to the block close to the timing disc (wrapped securely around a protruding water pump bolt works well) and bent so it comes over the edge of the disc and is aligned with the graduations on it.

Life can be made a little easier at this point if two of the flywheel bolts are partly threaded into the back of

Finding Number 1 piston TDC.

Using a dial gauge to find full lift on No.1 inlet lobe.

the crank opposite each other. When turning the engine back and forth, put a long screwdriver or bar between the bolts to apply a bit of leverage. Not that the beautifully assembled short motor should be difficult to turn over, it's just that it can get a bit stiff occasionally due to the engineering phenomena called stick-slip.

Now you need to accurately find Top Dead Centre for number one piston. With the piston visually at top dead centre in its bore, position the dial gauge stand on the top of the block and adjust it so the gauge is touching in the middle of the piston. This is necessary to limit any errors in the readings caused by the piston rocking in the bore. Piston TDC can now be established by turning the crankshaft back and forth and zeroing the dial gauge when the piston is at its highest point in the bore.

The trouble is this may not be the exact piston TDC position, depending on how accurate you wish to be. You may have noticed that while continuing

to turn the crank, the reading on the dial gauge remains at zero through a few more degrees of crank revolution, before the piston moves down the bore again and the reading changes. This is due to the piston dwell period at TDC, which is a function of crank stroke and con rod length, and so varies from engine to engine.

You can generally get a feeling for how long this dwell period is when turning the crankshaft to and fro and you need to set the degree wheel and pointer to zero at its mid point. However, if you wish to be deadly accurate then there's a bit more to do yet!

Rotate the crankshaft backwards through about one quarter of a revolution, then forwards until the dial gauge reads 0.001 inch (or 0.025mm) before the zero position. Make a note of the reading on the timing disc (say, 3 degrees before TDC). Continue turning the crankshaft forwards until the dial gauge moves off zero and reads 0.001 inch (0.025mm) again (say, 1 degree after

Half a thou before full lift, 99 degrees.

Half a thou after full lift, 110 degrees.

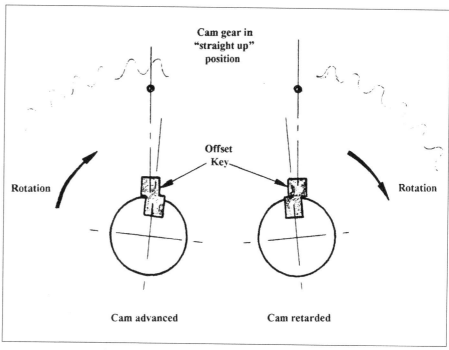

Schematic representation of effects of offset keys.

TDC). Make a note of this reading. Add the two readings together and divide by two (3 + 1 = 4/2 = 2). Carefully move the disc (backwards in this example) until the pointer is positioned at 2 degrees after TDC and tighten the nut. Then turn the crankshaft backwards one quarter of a turn, then forwards in order to re-check the timing disc is positioned properly with the pointer showing TDC. Tighten up the nut to clamp the disc securely – carefully so as not to move or dislodge anything. Double-check the readings and then this part is done.

A lightly oiled cam follower should be fitted into position in the block on the intake lobe of the number one cylinder (second hole back when viewed from the front). Slot in a pushrod from the top of the block and give it a wiggle to check it is seated in the follower correctly. The dial gauge and stand needs to be set up so that it seats in the little cup at the top of the pushrod, so the pushrod and the whole assembly stand vertically.

From now on always take any readings while turning the crankshaft forwards, in the direction of rotation. If you miss one or make a mistake, always turn the crankshaft backwards at least a quarter to half a turn before going forwards again, or the slack in the timing chain will cause no end of problems with the readings.

Slowly turn the crankshaft forwards until the camshaft reaches maximum lift and then zero the dial gauge. Now turn the crankshaft backwards about a quarter of a turn, then slowly forwards again until the dial gauge reads 0.0005 inch (or 0.01mm) before full lift, and it needs to be half a thou spot-on. Make a note of the number of degrees after TDC shown on the degree wheel. Slowly continue turning the crankshaft forwards, past full lift, until the dial gauge reads 0.0005 inch (or 0.01mm) after full lift, spot-on again. Make a note of the new reading on the degree wheel . Adding these two readings together and then

dividing the figure by two gives the number of degrees after TDC that full lift of the camshaft occurs. This can then be compared with the figure given on the specification sheet for that particular cam.

In the case of the standard 18V camshaft, for example, these are normally timed in at 106 degrees after TDC. The average of the two reading taken should be 106, if the cam timing is right. If it is within a degree or so either way then you are very fortunate as nothing more needs to be done, the cam is correctly positioned.

If, as is more often the case, however, the resulting full lift position is not what is wanted, steps must be taken to remedy the situation. If the number you have calculated is low ie 104, then the cam is advanced. If the number is high, say, 110, it is retarded. So to advance the cam it will have to be turned slightly forwards, in the direction it rotates and vice versa to retard it.

A selection of offset keys for the
camshaft sprocket.

5 degree key to advance camshaft 5
degrees.

With 5 degree key cam is now installed at
100 degrees full lift (half a thou before = 95
degrees, half a thou after = 105 degrees).

Vernier camshaft pulley.

When using the standard cam sprocket the timing can be changed by using an offset key in the cam keyway, in place of the standard one. These keys are available with offset increments from one to nine crank degrees (0.5 to 4.5 cam degrees – remember, 1 cam revolution takes two crank revolutions). Also, by moving the chain around by a single tooth on the cam sprocket giving an eighteen crank degree change, reversing the key allows increments back from the eighteen. The permutations are all there, they just require a little juggling to get the correct result.

Always re-check the timing again after making any adjustments to ensure that you didn't mistakenly put the key in the wrong way around.

By far the easiest means of altering the cam timing is by using an adjustable or vernier pulley. With the adjustable pulley the locking bolts are slackened allowing the cam to be turned without moving the crank, so you just have to set the crank to the correct number of degrees that full cam lift should be, and then turn the cam until the dial gauge shows full cam lift. Lock up the retainers and the job's done, bar a final check to ensure the timing is definitely correct. With the vernier it is a matter of putting the pin in another hole and re-checking the timing until correct.

Vernier or adjustable sprockets do make life easier, but cost considerably more than a set of offset keys. So you pays your money and takes your choice in terms of which to use.

Having said all this it must also be said that cam timing in practice is far easier than the long-winded description makes out and the accompanying photos should help make it clearer. Most people have learned how to time a cam from a book and, with a little bit of patience and thought, degreeing a cam is really quite straightforward and the end result very rewarding.

Appendix 2
Specialists & suppliers

Aldon Automotive
Breener Industrial Estate
Station Drive
Brierley Hill
West Midlands
DY5 3JZ
England
Tel: 01384 572553
Fax: 01384 480418
www.aldonauto.co.uk
Speciality distributors and electronic ignitions. Carburation and fuel supply, fuel injection. Tuning work, inc. rolling road.

Peter Burgess
Unit 1
Amber Buildings
Meadow lane
Alfreton
Derbyshire
DE55 7EZ
England
Tel: 01773 520021
Fax: 01773 520021

www.peter-burgess.com
Contact: Peter Burgess
Cylinder heads. Engines. Rolling road tuning. Valve guides.

Cambridge Motorsport Parts Ltd
Unit 5
Lacre Way
Letchworth Garden City
Hertfordshire
SG6 1NR
Tel: 01462 684300
Fax: 01462 684310
sales@cambridgemotorsport.com
Motorsport & performance parts.
JE Pistons.

Chemodex
www.chemodex.co.uk
Zincoat oil additive.

Hi-Gear Engineering Ltd
82 Chestnut Avenue
Mickleover
Derby

DE3 9FS
Tel/Fax: 01332 514503
www.hi-gearengineering.co.uk
Contact: Pete Gamble
5-speed gearbox conversions for classic MGs

Maniflow Ltd
Mitchell Road
Churchfields Ind. Est.
Salisbury
SP2 7PY
UK
Tel: 01722 335378
Fax: 01722 320834
www.maniflow.co.uk
Classic car exhausts and manifolds.

Mechspec MG Centre
Prospect Hill Farm
Gainsborough Road
Wiseton, Nr Doncaster
DN10 5AA
UK
Tel: 01777 818283

Fax: 01777 818299
www.mechspec.co.uk
Contact: Dave Parker
Excellent supplier of new and used parts.

Merton Motorsport
Merton Farm House
Dallinghoo
Noopbridge
Suffolk
IP13 OLE
England
Tel: 01473 737256
Fax: 01473 737798
Contact: Gerry Brown
Road, rally and race preparation.
Supplier of competition parts. Specialist
engine builder.

MG Motorsport
Wayside
Hempstead Road
Bovingdon
Herts
HP3 0HF
England
Tel: 01442 832019
Fax: 01442 832029
Contact: Doug Smith
Road and race preparation. Standard
and competition parts.

Moss Europe
Unit 16
Hampton Business Park
Bolney Way
Feltham
TW13 6DB

UK
Tel: 020 8867 2020
Fax: 020 8867 2030
www.moss-europe.co.uk
Everything! Special tuning performance
manual/catalogue also contains useful
tuning information.

Octarine Services
Octarine Services online
11 Byfield
Eastwood
Leigh on Sea
Essex
SS9 5TG
UK
Tel: 07801576731
www.octarine-services.co.uk
Contact: Chris Betson.

Other supplier addresses can be readily sourced from the internet, or from *MG Enthusiast* or
Safety Fast! magazines. Outside the UK, try national and local MG clubs.

Veloce *SpeedPro* books –

978-1-903706-59-6

978-1-903706-75-6

978-1-903706-76-3

978-1-903706-99-2

978-1-845840-21-1

978-1-787111-68-4

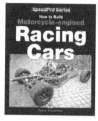

978-1-787111-69-1

978-1-787111-73-8

978-1-845841-87-4

978-1-845842-07-9

978-1-845842-08-6

978-1-845842-62-8

978-1-845842-89-5

978-1-845842-97-0

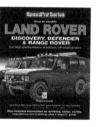

978-1-845843-15-1

978-1-845843-55-7

978-1-845844-33-2

978-1-845844-38-7

978-1-845844-83-7

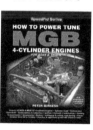

978-1-787113-41-1

978-1-845848-33-0

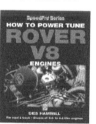

978-1-787111-76-9

978-1-845848-69-9

978-1-845849-60-3

978-1-845840-19-8

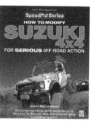

978-1-787110-92-2

978-1-787110-47-2

978-1-903706-94-7

978-1-787110-87-8

978-1-787111-79-0

978-1-787110-01-4

978-1-901295-26-9

978-1-787113-34-3

978-1-787110-91-5

978-1-787110-88-5

978-1-903706-78-7

ISBN: 978-1-787110-46-5

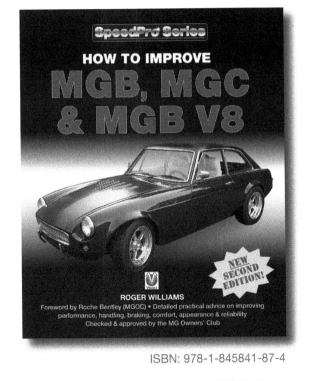

ISBN: 978-1-845841-87-4

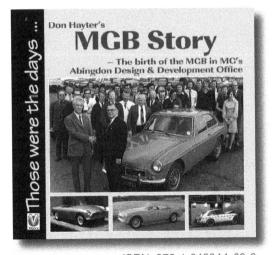

ISBN: 978-1-845844-60-8

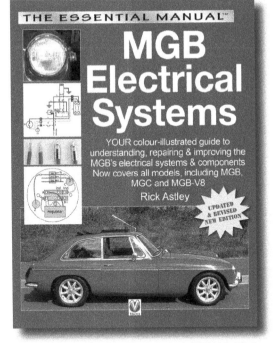

ISBN: 978-1-787110-52-6

ISBN: 978-1-787110-54-0

More great titles available from Veloce Publishing!

For more info and pricing information, visit our
website at www.veloce.co.uk
email: info@veloce.co.uk
Tel: +44(0)1305 260068

Index